Bountiful McKenzie

Bountiful McKenzie

The Story Of The Eugene Water & Electric Board

by

Norman F. Stone

Parkstone Company
Eugene, Oregon
U.S.A.

P.O. Box 5112
Eugene, OR 97405

Library of Congress Cataloging-in-Publication Data

Stone, Norman F., 1925–
Bountiful McKenzie

Includes index.
1. Eugene Water & Electric Board—History.
2. Water districts—Oregon—Eugene—History.
3. Electric utilities—Oregon—Eugene—History.
I. Title.
HD2767.O75E947 1985 363.6'0979531 85-61701
ISBN 0-9614776-0-1 (pbk.)

To those employees who gave their lives
in service to Eugene's municipal utility operation

Fay Tate, September 25, 1930
Joel E. Ballard, January 6, 1951
Kenneth E. Coleman, August 30, 1952
Arthur E. "Red" Barger, August 26, 1958
Rodney L. Franklin, January 3, 1967

CONTENTS

FOREWORD

IN 1986 EUGENE WATER & ELECTRIC BOARD will celebrate seventy-five years as a municipally owned provider of water and electricity. EWEB is Oregon's largest publicly owned utility and is well-known and respected throughout the Pacific Northwest for its general excellence as a public utility and for its strong leadership role in shaping responsible energy policy for the region.

A serious typhoid epidemic in 1906 intensified the movement toward city ownership of its water supply, and on April 18, 1908 a Charter amendment was approved by Eugene voters authorizing the sale of $300,000 in bonds and the establishment of a commission to be known as the Eugene Water Board. The existing waterworks system of the Willamette Valley Company was purchased in September of that year, followed by construction of a filter plant and the Walterville Hydroelectric plant which would produce electricity for pumping the water. On March 11, 1911 the entire plant was turned over to a five-member commission appointed by the City Council, to serve without pay and direct the management and operation of the municipal facility.

While Mayor J. D. Matlock was the leader of the movement for municipal ownership of the water and electric utilities, it was his predecessor, Mayor F. M. Wilkins, who suggested the creation of a permanent board, ". . . taking the whole matter from the hands of the council, and in this way place the future of the water system beyond all political strife." The autonomy of the Eugene Water Board fostered the development of an efficient, innovative and fiscally sound municipal utility system which has never required tax support. In fact, since 1946 Eugene's municipal utility has paid over $43 million in lieu of tax into the city's general fund and more than $16 million to school districts in its service territory.

The fascinating historical account which follows has been written by Norman F. Stone, a scholarly man of letters and history, who was part of EWEB for more than twenty-seven years. Stone came to EWEB in 1956 as a personnel officer, eventually becoming personnel manager and then, in 1971, director of administrative services. Following his retirement from EWEB in December 1983 Stone was able to devote full time to the massive research effort which is necessary when history is to be written. The result of his work is a thorough, honest and impressive review of EWEB's first seventy-five years.

As you read this history you will learn about many of the people and events that played a part in the evolution of EWEB from its valiant beginnings to its present leadership role among the public and private utilities of the Pacific Northwest. At the heart of EWEB's success are countless men and women—engineers, meter readers, commissioners, customer accounts representatives, general managers, drafting technicians, utility workers, conservation analysts, power dispatchers, cashiers, data entry operators, division directors, secretaries, linemen, accountants, to name but a few occupations represented at EWEB—men and women whose standards of excellence have been second to none. EWEB today, as throughout the past seventy-five years, has a strong commitment to the community it serves, and by continuing to provide excellent services at low cost, EWEB will play a key role in economic diversification and development in the Eugene-Springfield metropolitan area.

Camilla P. Pratt
President, EWEB Board of Commissioners

ACKNOWLEDGEMENTS

PROFESSOR SCOTT D. WALTON ESTABLISHED the format for *Bountiful McKenzie* in work that he performed for EWEB during the early 1960s. Professor Walton, then on the faculty of the School of Business, University of Oregon, drafted five chapters and had begun writing the sixth. He treated EWEB history within a framework based generally on the periods of administration of the various superintendents.

The work that Walton accomplished was excellent. It has been enlarged upon and revised somewhat to conform to the tone of the entire effort. But wherever possible, passages written by Walton remain intact.

After leaving Eugene and the University of Oregon, Walton joined the faculty at the University of Washington. He is now on the faculty at Ohio University, Athens, Ohio.

Anyone in this community pursuing historical material will at some point likely benefit from the help of Marty West, curator of the Lane County Historical Museum. West offers much more than her comprehensive files of information and photographs. She provides her encouragement as well. It is gratifying to know that there are those like Marty West in the community who are dedicated to the preservation of our history.

Indexing was provided by Marijoy Rubiloff, librarian at *The Register-Guard*. Rubiloff, like Marty West, has a profound interest in this community's history.

Many of the personnel at the University of Oregon Library and the Eugene Public Library provided ongoing help. I also appreciate the assistance I received from the State Archives office in Salem, University of Oregon Archivist Keith Richard and the Oregon State Library in Salem.

Oregon Public Utility Commissioner Gene Maudlin provided me access to a doctoral dissertation written by Carl Hoeber. This document outlines the history of regulation in the state and is a valuable resource.

My dear friends Murco and Margaret Ringnalda, long associated with Willamette University in Salem, have been a constant source of information and encouragement. They are masters of our language.

My longtime friend Professor Robert L. Scott, University of Hawaii, as did the Ringnaldas, provided his encouragement and advice.

Catherine Lauris, formerly with University of Oregon Books and one of Eugene's outstanding citizens, gave me valuable information regarding the acquisition of Library of Congress cataloging-in-publication data.

Glen and Helene Stadler, Mission Viejo, California, suggested the title of this book. Both, for many years, were important members of the Eugene-Springfield community. Glen Stadler served in the state Legislature, owned radio stations in this area and served for several years in a public information and lobbying capacity for the Eugene Water & Electric Board. Helene Stadler worked as a counselor at Lane Community College and later assisted her husband in legislative relations work in Salem. Glen Stadler is the author of *Courthouse*, published in 1984 by National Literary Guild, Inc. This fascinating, politically oriented novel draws upon Stadler's rich background in politics, both as practitioner and newsman-observer.

I appreciate the help given me by Edie Warner, Ruth Dellinger, Katie Solander, Patty Sabol, Helen Kershner and Vicki Maxon, all of the Eugene Water & Electric Board staff. I am also indebted to those over the years who had a special interest in the organization's history and with whom I held many discussions. These include Ford Northrop, Bill Eaton, George Easton, Leonard Kearney, Clarence Rutherford and Earle Clark. The comprehensive photographic coverage of the Carmen-Smith construction and activities in subsequent years is the work of the late Harry I. Gross.

Board of Commissioners President Camilla P. Pratt read page proofs on behalf of EWEB. Her comments and suggestions were of crucial importance. This is to say nothing of her discovery of at least three glaring errors in spelling.

Board Counsel Windsor Calkins also reviewed page proofs. Equally important was the help he gave me throughout the preparation of the text.

When I entered the Institute for Journalistic Studies at Stanford University in 1950, I noted that the *Eugene Register-Guard* was prominent among the American newspapers used as a subject of study by faculty and students. This may have been because Clifford F. Weigle, assistant director of the institute, had returned to Stanford after spending two years in Eugene as dean of the University of Oregon School of Journalism. But this connection notwithstanding, the *Guard* was in those times an outstanding newspaper and remains so today.

It was a fortunate day for Eugene when Alton F. Baker, Sr., in 1927, made his decision to leave the Cleveland *Plain Dealer* and make his home in this community. After purchasing the *Guard* in 1927, Baker brought William M. Tugman to Eugene from the *Plain Dealer* to be his managing editor. With the purchase of the *Morning Register* in 1930 the newspaper that emerged was the *Eugene Register-Guard*. The name remained unchanged until September 12, 1983 when the newspaper was redesignated *The Register-Guard* to more accurately reflect the regional nature of the publication.

I cite the *Guard* because of the enormous value of its files as an information source. It has also played a major role in this community's development. I would hope that at some future time, Warren C. Price's *The Eugene Register-Guard—A Citizen of its Community* will be updated. Volume I of this work was published in 1976, several years after Price's death. The book chronicles newspapering in this community dating from early times.

I also acknowledge Louis E. Polley's *A History of the Mohawk Valley and Early Lumbering,* published in 1984 by Polley Publishing. Lou Polley and I were colleagues at the Eugene Water & Electric Board for many years. His effort represents a significant and comprehensive contribution to the preservation of a part of this community's history. It also gave me a target. If my work proves to be half as valuable as Polley's, I will have achieved a major part of my purpose.

Norman F. Stone

CHAPTER ONE
The Years Before 1911

THE LATE FREDERICA COONS,* in her excellent *The Early History of Eugene*, a short guide for Eugene teachers, asks the question, "Why did people like the Skinners and Mr. Bristow leave their homes in the eastern and central states in 1843 . . . to come out here to settle?"

As Mrs. Coons said, reasons for undertaking the ordeal of crossing the plains varied among individuals. Elijah Bristow was a frontiersman and like Daniel Boone, enjoyed new country and the role of a pioneer. One reason that Eugene Skinner undertook the journey was his poor health, which he hoped might be benefited by a change in location. Perhaps the loss of their first three children, all daughters, at a young age in Illinois also influenced the Skinners in their decision. And of course, a reason common to all emigrants of that era was the promise of free land.

The history of Eugene during its early years is similar to that of hundreds of cities in the American West. The founders of these communities, which have become major cities in many instances, were people of strength, character and vision. While the names of the men are recorded most notably in historical accounts, it should be emphasized that the frontier woman possessed remarkable strength and contributed much to her community. Mary Skinner, who arrived with a young child, was to bear four more children, a total of eight in her lifetime including the three who were lost in Illinois.

In August, 1851 the town site was laid out between the Willamette River and Seventh Street. There were seven streets running east and west, numbered First through Seventh, and four running north and south, named Ferry, Mill, High and Water. When Mary Skinner was

* Mrs. Coons, a prominent Oregon author, was the wife of retired EWEB employee Loyd Coons. Both are deceased.

asked to name the town, she chose "Eugene City" in honor of her husband. In 1889 the name of the town was redesignated "Eugene."

By the turn of the century Eugene was a thriving community. The University of Oregon, which had its first enrollment of students in 1876, was well-established. The newspapers of the day reflected a general optimism and it was evident that the political, social and economic life of the community was well-developed. The forest products industry, agriculture, mining and merchandising activity in the city of Eugene all contributed to economic stability. In 1905 the population was reported as 5,743 in the city of Eugene, with a Lane County population of 23,740.

The first franchise for an electric system in Eugene had been authorized in an ordinance approved by the Common Council on February 14, 1887—seven years after the nation's first commercial electric system had been installed in the East by the Thomas Edison Company. The franchise was issued to H. C. Humphrey, J. F. Robinson, Wm. Edris and A. Klein and associates, including F. B. Dunn and J. C. Church. The firm's name was established as the Eugene Electric Company.

The generating plant, a 100-horsepower, direct-current type was located near what would have been the intersection of Seventh and Ferry streets. In 1897 the plant capacity was increased to 240-horsepower to meet increasing demand.

In 1903 ground was broken for a new plant in Springfield and the capacity would be increased to 1000-horsepower. Because there were customers in Springfield as well as Eugene, the name of the firm was changed to the more inclusive Lane County Electric Company.

On August 1, 1905 the electric system, including the power plant in Springfield and a substation in Eugene, was sold to the Willamette Valley Company, a subsidiary of an eastern syndicate known as Rhodes, Sinkler and Butcher. Stockholders in the Lane County Electric Company at the time of transfer included D. A. Paine, J. F. Robinson, F. W. Osburn, W. E. Brown, Dr. O. E. Smith, B. D. Paine, Booth-Kelly Lumber Company, Mrs. Abbie Smith, S. H. Friendly, Mrs. Sarah Lauer, S. S. Spencer, J. E. Yarnell, James Magladry, Mrs. Theresa Jackson and Ralph D. Robinson.

On February 10, 1886 the Common Council of Eugene had granted its first water-supply franchise to T. W. Shelton and Charles Lauer and Associates, including T. G. Hendricks, S. H. Friendly, J. C. Church, S. B. Eakin and J. F. Robinson. The name of the firm was the Eugene Water Company. The first source of Eugene's water supply was reported to be at the northeast end of Skinners Butte on the Willamette River.

Later principals of the firm included G. R. Chrisman and F. L. Chambers and when the water system was sold to the eastern syndicate, Rhodes, Sinkler and Butcher in 1905, Dr. T. W. Harris was president, R. McMurphey secretary-manager, S. H. Friendly treasurer and J. F. Robinson, director and stockholder.

EARLY INTEREST SHOWN IN MUNICIPAL OWNERSHIP

The ownership by an eastern syndicate of both the water and electric systems gave impetus to the cause of municipal ownership. Such remote ownership of vital community services, concerns about health and a strong sense of civic pride among Eugene citizens all contributed to shaping the future of Eugene's utility system.

Although Eugene was being supplied with water and electricity at the turn of the century, there were individuals and groups dissatisfied with the service provided by the two companies; they thought the city should enter into the utility business. These proponents of municipal enterprise eventually secured favorable action of the City Council, the State Legislature and the citizens of Eugene. The story of municipal-utility ownership in Eugene, up until 1911, describes how these challenges were met.

An indication of the problems relating to water supply was illustrated by a notice that appeared in the local press during mid-July, 1905. It read as follows:

To Water Consumers

The irrigation hours have been slightly changed, and are now as follows: 6 to 8 o'clock in the morning and 6 to 9 o'clock in the evening.

Sprinkling of streets through hose is absolutely prohibited.

Irrigation through an open hose, that is, through a hose without a nozzle or sprinkler thereto attached, is not permissible under any circumstances.

The above regulations are essential and are not at all unreasonable, and we must insist upon their observance, and that they may be enforced we will at once put on a few inspectors with instructions to turn the water off of any premises where a violation of any one of the above rules is discovered, and a penalty of $1 for each offense will be exacted before the water will be turned on again.

(Courtesy University of Oregon Library)

STRONG LEADERSHIP EMERGES

Early in 1904 promoters of municipal ownership were able to get the matter of the city acquiring its own electric and water plants before the voters. On April 4 one of the questions submitted to the electorate was, "shall the City Council take steps toward ultimately acquiring for the city its own electric plant and its own water plant." The proposition was advisory in nature with voting on the issue separated into questions concerning the two types of utility. Both a municipal light system and a city water system were approved by wide margins (for lights, 892; against, 133; for water, 910; against 116).

The prime movers in the push for muncipally owned utilities were members of the Fire and Water Committee of the Council. The committee consisted of Francis M. Wilkins, Eugene pharmacist, G. W. Griffith and W. W. Calkins, attorney and president of a local bank. The committee felt the existing service to be inadequate and too expensive. It was not thought that the Eugene Water Company was able or willing to extend the necessary mains or to provide a high enough pressure to accomplish the job of fire protection. Regarding electricity, the City Council had not ordered any new street lights since 1902 on the stated grounds that the price charged for electricity was too high. Favoring municipal ownership and believing that the privately owned utilities were not doing an adequate job, these council members made the following statement in support of the affirmative on the April, 1904 ballot:

> The second question is submitted at this time in order that the Common Council may have ample time to acquire light and water plants if such course is thought desirable. Owing to outstanding contracts, it will be several years before any action can be taken. Furthermore, our City Charter does not permit the city to issue sufficient bonds to install either a light or water plant, and it will be necessary to have it amended by the State Legislature before further action can be taken. It is therefore thought desirable to submit the question to a vote at this time.

GOVERNMENTAL PROCESS SET IN MOTION

There was also substantial support for municipal ownership by other council members and prominent citizens of Eugene. Joseph D. Matlock, farmer, businessman, former mayor and in 1904 a member of the Council, served on the health committee and actively supported the movement. The eminent jurist Joshua J. Walton lent his support, as did Martin Svarverud, a businessman and realtor, and later a member of the first Water Board.

With the approval of the voters, the Council set about securing the necessary changes in the City Charter.

On December 26, 1904 the Council, after discussing some needed charter amendments, appointed a committee to draft other necessary

Francis M. Wilkins (left), prominent, sometimes controversial, Eugene leader. Joseph D. Matlock (right), for whom Eugene power plant at Walterville was named. (Both photos courtesy Lane County Historical Museum.)

W. W. Calkins (left), member of the City Council and the Council's Fire and Water Committee which recognized need for municipal ownership of utilities. Calkins, an attorney, was an uncle of Windsor Calkins, present-day EWEB Counsel. (Photo courtesy Windsor Calkins). Martin Svarverud (right) later a member of the first Water Board was a strong municipal-ownership advocate prior to formation of the Board. (Picture courtesy Eugene Public Library)

charter changes. Early in January, 1905 the entire Council debated a proposal by this committee "giving the Council power to construct and maintain water works, electric lights and gas plants, and other public utilities, or empowering them to purchase those already established." This was approved. Appropriate proposals were drafted by the city attorney and the proposed amendment was ordered sent to Representative D. W. Griffin, member of the Oregon State Legislature from Lane County, to have him introduce it in the Legislature and to "father the bill."

Representative Griffin's effort brought forth an amended charter filed in the office of the secretary of state on February 18, 1905. Numerous provisions were included to enable the city to go into the water, gas and electric-light business. To enable the municipality to engage in such enterprise, some provisions were repetitious, perhaps in the expectation that opposition might result in the elimination of one or two sections, but with luck, not all. The main authority was contained in Chapter 10 of the charter entitled: "Of Water Works, Electric Lights and Sewers." The Common Council was therein authorized to construct complete systems of electric lights, gas lights and waterworks or—in lieu of constructing such systems anew—to purchase existing waterworks and gas and electric-light plants. The final step in securing adequate authority was taken on April 11, 1905 when the citizens of Eugene approved the charter amendment favoring the establishment of publicly owned water, gas and electric light systems.

Also as a result of the election of April, 1905, Councilman Wilkins became mayor of Eugene and he quickly took action. In his first speech as ex officio president of the Council, he said, "First I would recommend that the Committee on Fire and Water take up the question of municipal ownership of lights and water . . ." Mayor Wilkins' post on this committee was filled by Mr. Matlock, an even more ardent supporter of public ownership than Wilkins. Favorable committee and Council action brought the advocates of the municipal systems of water and lights to the next major hurdle—the securing of adequate financing to construct or purchase such systems. The matter was to be submitted to the voters of Eugene on September 11, 1905 when they would be asked to approve a $200,000 bond issue—$100,000 each for the water and electric systems.

On September 9, 1905 the *Eugene Morning Register* carried a "municipal ownership discussion." Editorially, the *Register* appeared to be neutral about the upcoming bond election on September 11, but warned "we should remember that if bonds are voted and issued, they must be paid. If the proposed indebtedness is added to what we already have, there will be represented an indebtedness of over $50 for every man, woman and child in the city. This is a point that should receive careful thought." Notwithstanding the *Register's* ambiguous position, the bonds were approved by the voters; for light bonds 504, against 163; for water bonds 501, against 153. The way seemed clear to proceed with the acquisition of physical properties.

PROCEDURAL ERROR BRINGS FRUSTRATION

The great expectations of those promoting municipal power and water met with frustration when the opponents, led by the management of the privately owned utilities, filed for an injunction charging that the September 11 election was illegal, as only one polling place had been provided when the City Charter specified that there should be four. Rather than consume time fighting the issue in the courts, the matter was again submitted to the electorate on January 2, 1906 with the letter of the law more closely observed. Arguing against the bonds, the *Eugene Daily Guard* on December 30, 1905 said, "And bigger taxes? That would mean depreciation of all real property, the little home, the costly home, business property alike. It would mean that men with money would stay away—would not invest in a tax-eaten town." On the day of the election the *Guard* said that, "While the bonds may carry, it is predicted by this paper that the plants will never be erected. Litigation will begin and probably will result in the abandonment of the enterprise." The opponents of municipal ownership were in large part successful. The light bonds were defeated by the vote of 488 for and 539 against. The water bonds passed by a scant margin of eight votes: 500 for; 492 against. This dramatic change of the voting in the two elections probably was brought about by a big increase in the "no" votes with little change in the number voting "yes." In a stormy Council session following this rebuff to public ownership, the bond issue for the waterworks was decided against and the matter permitted to rest for the time being.

EPIDEMIC SPURS MOVEMENT

Almost as soon as the decision was made not to construct or purchase a water system, the cause of public ownership suddenly gained unexpected support. Ironically, on July 11, 1905 at a meeting of the City Council, it was reported that, "the city was . . . free from contagious disease." During the late summer of 1905 newspaper columns in Eugene carried stories about a disastrous yellow fever epidemic in the southern United States. This event seemed far away and of little danger to the comfortable community of Eugene. However, in January 1906 typhoid fever struck and Eugene experienced what the State Board of Health then characterized as the "worst epidemic in the history of Oregon." When the city's water supply was found to be the cause of a typhoid epidemic, public indignation with the private water company mounted. On February 10, 1906 the *Guard* headlines were as follows:

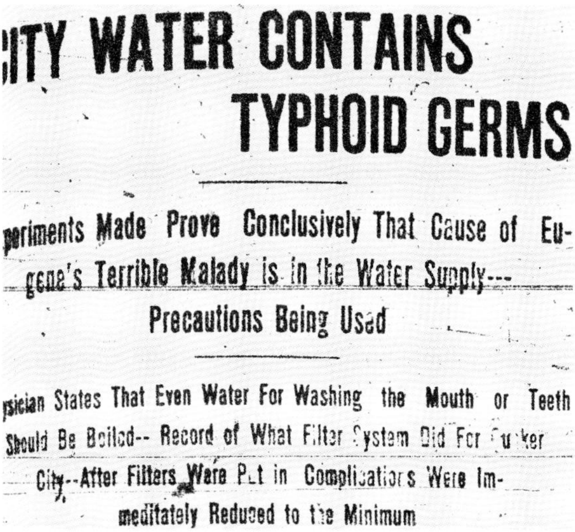

ITY WATER CONTAINS TYPHOID GERMS

periments Made Prove Conclusively That Cause of Eugene's Terrible Malady is in the Water Supply--- Precautions Being Used

ysician States That Even Water For Washing the Mouth or Teeth Should Be Boiled-- Record of What Filter System Did For [illegible] City--After Filters Were Put in Complications Were Immediately Reduced to the Minimum

(Courtesy University of Oregon Library)

Although medical investigation was not well-advanced in 1906, it was determined that only those drinking water supplied by the privately owned system contracted the disease. Those who obtained their water for household use from other sources were spared. The Willamette Valley Company, a subsidiary of an eastern syndicate which had acquired ownership of the Eugene Water Company in the summer of 1905, issued a statement that it had had actual possession of the waterworks only since February 1, 1906 and that it was doing all that was possible to remedy the situation. The public was not satisfied and Mayor Wilkins was asked to resign at a mass meeting of citizens on February 21, 1906 when it was declared that he had not done all in his power to clean up the typhoid menace and that he had "turned his coat" on the matter of public ownership. The mayor denied these charges, but expressed doubts as to the wisdom of launching a municipal-utility program until the new company had had a chance to prove itself.

On February 23, 1906 the *Daily Guard* carried the following headline:

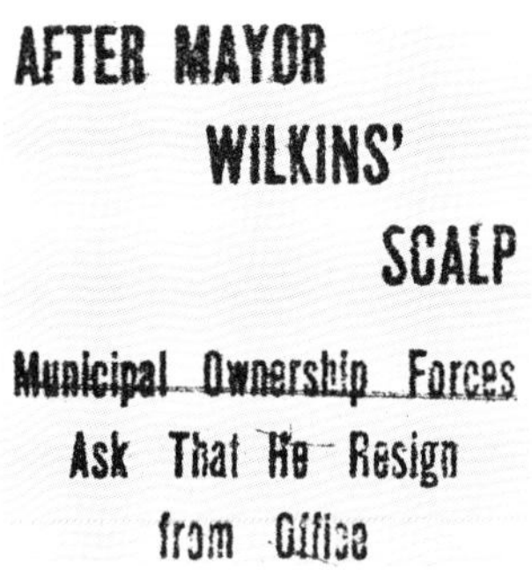

AFTER MAYOR WILKINS' SCALP

Municipal Ownership Forces Ask That He Resign from Office

(Courtesy University of Oregon Library)

The Oregon State Board of Health does not have records of Eugene's typhoid epidemic. However, in a byline article appearing in the *Eugene Register-Guard* in 1956, longtime Eugene resident Blanche M. Thurston said, "many deaths occurred in 1906." Old-timers also remembered that many women in Eugene who survived the disease suffered such a loss of hair that bobbing became fashionable for a short time.

No immediate public action followed the typhoid epidemic, but 1907 witnessed the beginning of another chain of events which finally resulted in the city's being in the utility business. Matlock, who had replaced Wilkins on the Fire and Water Committee, replaced him as mayor in the April, 1907 election. Candidate Matlock, a Eugene merchant and former mayor, based his 1907 election campaign on proposals to place the city utilities under municipal ownership. One month after his election conferences were begun by city officials with the water company manager to determine whether or not the private company was willing to sell its properties and, if so, what a "fair price" would be. The negotiations culminated in the water firm's proposal to sell certain of its properties to the city for the sum of $140,000.

The Council recognized that greater demands would be made for water than could be met with the existing facilities and therefore proposed to the citizens of Eugene that they authorize the issuance of bonds up to $300,000 for the purpose of acquiring the existing facilities and constructing necessary additions. In the election the water bonds were approved: 567 for; 171 against. Authorization to purchase the water plant was approved: 635 for; 125 against.

WATER OF PRIMARY CONCERN

It was initially proposed to supplement the existing facilities for providing water to the people of Eugene with a gravity water system, with water to be piped from Ritchie Creek, a tributary of the McKenzie River. A survey made early in the summer of 1908 disclosed that this would be quite costly. Since more than the remaining $160,000 of the $300,000 bond issue would be needed to add the gravity system, the Council proposed that the authorized bond issue be increased to $500,000. This was presented to the voters in a special election held October 1, 1908 and was turned down. The vote was for, 332; against, 490. As a result, the gravity project was abandoned and the Council was then forced to turn to a less costly pumping system. Its decision to provide its own power for the pumps ultimately established the city in the electric-utility business.

MUNICIPAL OWNERSHIP BECOMES REALITY

On November 18, 1908 the City Council acquired the water properties of the Willamette Valley Company in exchange for proceeds from the sale of $140,000 of bonds which it was authorized to issue. The remainder of the $300,000 bond issue was sold and part of the proceeds were

In the early part of the century the millrace was the center of Eugene's commercial activity. (Upper photo) Eugene Flour Mill. (Lower photo) Eugene Woolen Mills office on left, Eugene Fruit Growers Plant on right. (Photos courtesy Lane County Historical Museum)

Eugene Mill Company, east end of Fifth Street. (Courtesy Lane County Historical Museum)

used to obtain a site for the power plant and to secure rights-of-way for a four-mile canal (to bring water to the power plant from a sufficient height for power generation) and for transmission lines to Eugene. The power plant was to be constructed on the McKenzie River near Walterville and power would be transmitted to pumps in Eugene near the Willamette River. It was the intent of the Council to turn over the operation of the water system to a Board as Mayor Wilkins had suggested in 1906.

The proposal submitted to the voters in April of 1908 had not only asked for approval of a bond issue to secure a water plant, but also stated "the income of such plants in operation, maintenance, extension and improvement thereof after said plant shall be so completed as to be ready for operation shall be in the hands of a commission to be known as the Eugene Water Board." After the water system was purchased it was proposed at the Council meeting on November 25, 1908 that the Board be appointed. City Attorney L.E. Bean advised, however, that it would be premature to do so since the complete plant was not ready for operation. Therefore the Council and mayor were obliged to manage the affairs of the water system until the Walterville project was complete and power forthcoming to operate the new pumps.

The Council hired an engineer, Frank C. Kelsey, to plan and construct the power plant near Walterville and employed S. W. Taylor (who had

Matlock Station (upper photo) 8-ft. diameter penstocks from flume to power house. (Lower photo) Forebay and power house. (1912 Annual Report, Eugene Water Board)

Eugene filter plant and substation. (1912 Annual Report, Eugene Water Board)

been superintendent for the Eugene Water Company since 1890) to be superintendent of the city-owned water system. Still the City Council was obliged to meet frequently, sometimes five or six nights in a single month, in order to carry out the utility program while also transacting other business of the city.

In the beginning the water system was a financial success. In the first four months of municipal ownership and operation, it earned slightly over $1,000. By April of 1910 it was earning almost $2,000 per month over its expenses according to a report made to the City Council. Although it had been provided in the 1908 authorization that some of the income might be employed in making extensions and improvements to the existing system, there was not enough money to completely finance the construction of a filtration plant that the Council decided to build in 1910, nor were there sufficient uncommitted funds from the bond issue of 1908. The Council therefore called a special election to ask for more money for this purpose. On May 16, 1910 the voters approved an additional bond issue of $60,000, of which $40,000 was to be used for the completion of the filtration plant and the remainder for other water-system improvements. The vote was: 360 for; 252 against. By the end of the year, construction of the power plant was almost completed and that of the filtration plant was well underway. The superintendent of the water department, S. W. Taylor, resigned on December 8, 1910 and City Councilman R. B. Henderson, who had served long on the Fire and Water Committee, resigned his Council post and assumed the superintendency.

THE BEGINNING OF AN ERA

Early in 1911 the Walterville Power Plant was completed. February 17, 1911 was the day of the "open house" celebration at the plant with a large gathering of city officials and citizens in attendance. Among them was Mayor Matlock, for whom the plant was named in recognition of his devotion to the cause of municipal ownership. With the Walterville Plant in operation and the filtration plant near completion, the City Council decided to turn the operation of the utility over to a water board as had been provided in the charter amendment of 1908. Accordingly, the Council passed an ordinance to implement the charter provision, naming E. U. Lee, M. Svarverud, C. S. Frank, R. M. Day, and B. B. McKinney members of a commission to be known as the Eugene Water Board. With the completion of construction and acquisition of plants and the passing of control of the utility into the hands of a group other than the Council, an end was made to the beginning and a new era entered upon.

CHAPTER TWO

The Early Years - The Eugene Water Board

IN THE EARLY YEARS of operation of the Eugene Water Board the municipal utility was confronted with many problems and the successful meeting of the challenges is a testimonial to the wisdom and bold action of that body. With little provision made in the charter and ordinance for the administration of the utilities, the Board was confronted with difficulties in the management of the enterprise. With surplus electrical energy at its disposal, the issue of its use was raised. Financial questions faced the young organization and it struggled to find answers. As the operations of the Water Board increased, problems in relation to the community mounted correspondingly.

BOARD TO HAVE "ENTIRE CONTROL ..."

On February 20, 1911, the City Council passed an ordinance creating the Eugene Water Board. It was signed by Mayor Matlock on February 27 and on March 11, the first Board meeting was held. R. M. Day was made temporary chairman and E. U. Lee temporary secretary. Day was elected president and Martin Svarverud vice president. As provided for by the ordinance adopted February 20, lots were drawn which gave Day the one-year term, Svarverud two years, Lee three years, C. S. Frank four years and B. B. McKinney the five-year term.

The new Board was composed of freeholders serving without pay and without very specific instruction from the City Council as to the manner of operation. It was to have "entire control of the water plant of the City of Eugene, including the city power plant. . . ." and its duties were "to operate the entire system and make extensions to the same when deemed advisable, and have entire control of all water funds and disbursements thereof and . . . keep an account of the same . . ."

WHO WAS FIRST SUPERINTENDENT?

The relationship between the Board and the actual supervision of operations during the early times is unclear. Just three days after its first meeting, the Board met again and authorized Alvin Meyers "to telegraph Fred M. Blackburn that the Board will pay him $2,000 per year to take the superintendency of the electric plant and waterworks of Eugene, Oregon, providing that said Blackburn fills the said position to the entire satisfaction of the Board, the Board reserving the right to discharge said superintendent at any time it sees fit." Meyers, who had been employed by the City Council on April 18, 1910 to act as general superintendent of the electric system, took charge of electrical operations under the new Board and R. B. Henderson continued as water superintendent. It is not known what role Blackburn played in his relationship to Meyers and Henderson. One account indicates that Blackburn was "mechanical superintendent" of the operation. The Board minutes reveal that Blackburn was on the payroll through September, 1911. Those same minutes, documenting the Board meeting of October 9, 1911, make reference to Superintendent Meyer (sic). It appears that Meyers left the organization early in 1912, and at its meeting on May 9, 1912 the Board approved a motion to send a telegram to Meyers requesting that he "return and resume duties as General Superintendent."

It is generally accepted by other observers that Alvin Meyers was the first general superintendent of the Eugene Water Board, but the matter

c. 1890; Eugene City Hall. First Water Board meeting, March 11, 1911, was held in "water office" at the building. (Courtesy Lane County Historical Museum)

remains unclear. In the April 2, 1911 edition of the *Eugene Morning Register* Blackburn is referred to as the "Utilities Superintendent" and the "cultured superintendent of the water and power systems." Later, on April 11, 1911, Blackburn and E. U. Lee were instructed to place in execution a system of accounting for the Eugene Water Board suggesting that Blackburn was engaging in activities other than those of "mechanical superintendent."

We take the position that Blackburn was the first chief executive officer of the Water Board. His tenure was brief and perhaps the new Board soon found reasons to replace him. But at least for a short time, he was recognized by the community as the Board's chosen superintendent.

Meyers returned in June, 1912, but again resigned in June, 1913 because of the serious illness of Mrs. Meyers and the necessity that she "be taken to a different clime in order to reclaim her health." Cyrus Avery Whipple was employed as general superintendent effective July 1, 1913. However on July 13, 1914 he presented his resignation "due to sickness of one of the members of his family." C. W. Geller, who had been salaried secretary to the Board since May, 1912, was appointed acting general superintendent on July 14, 1914 and on October 4, 1915, the Board adopted a motion that "Mr. Geller be hereby appointed general superintendent in addition to his duties as secretary, and that his title henceforth be known as General Superintendent-Secretary . . ." Geller served until 1918, the first relatively permanent occupant of the office.

ONLY TWO ELECTRIC CUSTOMERS THE FIRST YEAR

The instability in the chief administrator's office did not prevent bold action on the part of the organization. The Board was composed of dedicated men who were willing to spend long hours in discussion of means to make the utility successful. Through 1916, only one change took place in its composition. J. W. McArthur, who had been associated with the utilities in a number of capacities, defeated Martin Svarverud in the election of 1913. This brought no important change in policy thinking to the Board, however, as both men were devoted to the cause of public ownership of the city's utilities. McArthur, in his excellent "Reminiscences" in the 1946 Annual Report, writes, "M. Svarverud, the man I defeated, was a real patriot too, and a staunch supporter of municipal ownership. Though defeated for a position he was proud to hold, he never held it against me and remained a friend all his later life, and was always ready to help further the success of the utilities."

During 1911 the electric system acquired only two customers, one of which was the Eugene Planing Mill. It should be remembered that the electric utility was still in competition with the Oregon Power Company. Records indicate that by the time the 1912 Annual Report was published in December of that year, there were 1,800 customers on the water system.

WATER TREATMENT BASICALLY SAME AS TODAY

The water system intake was located on the south bank of the Willamette River about 300 feet east of the pumping plant. The pumping plant was located at the present-day site of the EWEB pole and pipe yard. Water was transmitted from the intake to the pumping plant by means of a 12-inch steel pipe. Its capacity at high river stage was three million gallons per day.

Both aluminum sulfate and chlorine were used in the settling and filtering process, just as is the case today. Some 30 miles of mains existed in 1912, some wooden. Testing of the water was conducted by Professors O. F. Stafford and A. R. Sweetser of the University of Oregon and the 1912 report states that since the filters were installed "no case of typhoid fever or other ailment has been traceable to the use of city water."

By the time of the 1912 Annual Report, Alvin Meyers was established as general superintendent. The office organization included Meyers (upper-middle). Others (left to right, front row) were C. W. Geller, paid-secretary to the Board; Verna Clubb, chief accountant; Violet Peart, cashier; and F. W. Dietrich, assistant engineer. (Upper oval picture left) E. W. Rand, superintendent of power plant; (upper oval picture right) C. E. Russell, pumping plant superintendent. Flanking Meyers: (left) W. S. Lyons, water department inspector (right) O. M. Douglas, electric distribution superintendent. (EWEB files)

CAREFUL ACCOUNTING METHODS BEGUN EARLY

There were two reservoirs located on Skinners Butte, only one of which was suitable for domestic purposes. The larger reservoir, located on the west end of the butte, had a capacity of two million gallons and was used for fire protection only. The smaller reservoir, with a capacity of 225,000 gallons, was located at the east end of the butte. An eight-inch pipe carried water from this reservoir down the south side of the butte and along Willamette Street to Ninth Street.

Property owners desiring a water main signed a petition listing the names and addresses of the petitioners and the amount of annual revenue they could guarantee. It was possible to closely estimate revenues from data on mains laid previously. In addition to estimating revenues based upon accounting experience, the Water Board physically inspected the property involved before acting.

No mention of water rates is made in the Board minutes of 1911. Therefore, it can be assumed that the rates that had been established by the City Council on January 31, 1910 remained in effect.

(Left) Alvin Meyers, thought by early observers to be first general superintendent although local press recognized Fred M. Blackburn as holding the position for a short time after the first Board took office. (Right) G. W. Geller who was appointed acting general superintendent July 14, 1914 and given a permanent appointment as general superintendent-secretary on October 4, 1915. Geller, who served until 1918 was the first relatively permanent occupant of the chief executive's office. (EWEB files)

BOARD DEMONSTRATED CONCERN FOR EQUITABLE RATES

Ordinance No. 960 set forth comprehensive water rates. The basic charge for a dwelling was 50 cents per month with an additional charge of 10 cents for each person exceeding eight. The bathtub, water closet and urinal charge was 25 cents for each fixture. The charge for a steam or hot water heater was 50 cents. Lawn sprinkling was at a rate of five mills per square yard and barber shops were charged 75 cents for the first chair and 25 cents for each additional chair.

The metered water rate appeared to be 18.7 cents per 100 cubic feet. The 1912 Annual Report states that 26.2 percent of the Board's water customers were metered, pointing out that this customer class produced only 20.6 percent of water revenues. Notwithstanding this apparent disparity in the water rate structure, the Board recommended to the City Council, as part of its request for a $100,000 bond election, that $15,000 be expended on additional water meters. The report stated, "The amount for meters is recommended without anticipating any net gain financially, but for the sake of justice and equity. Equitable charges and pure water are considered a duty above mere financial gain."

Accompanying the rates were extensive rules and regulations which stated the conditions for receiving service. The rules prohibited supplying water to others. A new application was required for each additional faucet installed. Special application was required for yard sprinkling during June, July, August and September while no application or charge was required for the other months of the year.

Street sprinkling was prohibited and only the sidewalk in front of the dwelling served could be cleaned by water. The use of hose when alarm of fire was sounded was absolutely prohibited except to protect property.

An additional ordinance, No. 961, provided for fines and imprisonment for violations of water rules and regulations and for failure to notify the city waterworks when new fixtures were added.

THREE SOURCES OF POWER

In the early years power was supplied to the city from three sources. These were the Chambers Power Company, which had a generating facility on the Willamette River, the Oregon Power Company, which generated power in its Springfield plant and the municipal-hydroelectric power development near Walterville.

The Chambers Power Company incorporated February 2, 1906 to "erect, operate and maintain plants for the production of electricity for lighting, power and other uses . . ." The principals were F. L. Chambers, George Midgley and Darwin Bristow. The company built a low-head dam on the Willamette River just above Interstate 5, the remains of which are still visible, and was associated with the Eugene millrace. It served such customers as the Eugene Woolen Mill and the Eugene Excelsior Plant.

While the Water Board was to move aggressively against competition from the Oregon Power Company, no reference is found to concerns about the Chambers Power Company. Chambers was a prominent citizen, active in business and banking affairs. It is possible that the Board was content not to interfere with the Chambers Power Company and perhaps thought that the Chambers operation would eventually be absorbed by the municipal system. This occurred when the company was dissolved in 1934.

The intake for the Walterville Plant, or Matlock Station as it was known in those days, was located on the north bank of the river 18 miles east of Eugene. Two generating units were in service at the plant, each with a normal rating of 880 horsepower. Energy was transmitted at 23,000 volts over 15.5 miles of single line to the city substation at Eighth Street and the Southern Pacific Railway. Power distributed from the city substation was used for pumping city water, street lighting, and commercial light and power service. One circuit served the north and west portions of the city and the other the south and east portions. This was said to permit continuous extensions. It was the plan of the city of Eugene to place at least one 100-watt series tungsten lamp at every block corner. The 1912 report states that 30 miles of city streets were lighted by this system. The number of electric system customers rose to 492 in 1912.

Interior of Eugene sub-station, c. 1915. (EWEB files)

BOARD MOVES TOWARD ELECTRIC UTILITY OWNERSHIP

Among the few specific duties given the Board by the Council in 1911 was "to sell the public at uniform rates to be fixed by it, any surplus electrical power over and above what may be needed for the operation of the water plant or any other municipal purpose." The charter amendment of April, 1908 made no reference to the city entering the electric utility business. But we recall that when the voters later that year turned down a proposal to increase the bond issue in order to provide for a gravity system of water supply, the Council may have foreseen the opportunity to go into the electric business. The fact that a plant was constructed to generate electricity considerably in excess of the immediate and prospective needs for pumping water gives support to the concept that city leaders had been planning to enter this field of enterprise even without new authority from the citizens. In any event, the Board took firm steps to become the sole distributor of electrical energy in Eugene.

FIRST ELECTRICAL RATES COMPETITIVE

The first move was to secure the necessary financial resources. On May 8, 1911 the Board requested the Council to put new charter amendments before the people in order that additional funds might be secured, including $25,000 "for the purpose of installing a Light Distributing System . . ." At the special election held on July 11 of that year, the voters authorized the Council to issue "light, power and water bonds," part of which were to be used to secure funds for adding to the lighting system and to improve the power plant. The vote was 633 for; 355 against.

On November 16, 1911 a motion was adopted containing rates, rules and regulations governing the sale of electrical energy, and the Eugene Water Board was now engaged in a new enterprise. The electric rates established by the Board on November 16, 1911 indicated that it was aggressively carrying the fight to its competition. The Oregon Power Company, owned by the eastern group which had purchased the local water, gas and electric utilities in 1905, had charged rates ranging downward from 15 cents per kilowatt-hour. The first schedule of rates for "lighting service" published by the Eugene Water Board set a maximum of nine cents per kilowatt-hour for the first 20 kilowatt-hours, with a minimum of three cents per kilowatt-hour for all energy consumed in excess of 1,300 kilowatt-hours per month. A minimum charge of $1.00 was established. Power service was at an even lower rate, with five cents per kilowatt-hour being charged for the first 100 kilowatt-hours used per month and two cents for all over 10,000 kilowatt-hours. A demand charge was imposed amounting to 75 cents for each horsepower of maximum demand. In its report to the Council dated January 1, 1912, the Board reported progress in the extension of the lighting distribution system and that the rates were based "on the presumption that the citizens will patronize their own light and power system. Should they fail to do so, there will necessarily be a deficit each year which the taxpayers will be compelled to pay."

There were some special rates made during the first year of the sale of electrical energy, both for competitive reasons and as a public service. In

February, 1912 the general superintendent was authorized to accept contracts for electrical energy supplied to dealers or firms whose principal business was the sale of electrical energy-consuming devices and apparatus at the rate of four cents per kilowatt-hour. At the same time, a special monthly rate was made for residents who wished to leave their porch lights burning all night as a safety measure. However, by 1915 the policy had firmly been set that the Board would not permit prospective customers to secure exceptionally favorable rates by threatening to place business with the competition. The question of furnishing electric energy to the new armory had come up, and the state military department had indicated that it would place its electric contract with the Eugene Water Board if it was assured that any competitive rate would be met. The Board responded that "we could not deviate from our usual practice and policies which have been previously set, that the rates be the same to all consumers in proportion to use, and that we could not guarantee to meet any competitive condition unless same was given to all of our similar customers."

Canal intake above Walterville, c. 1910.

In 1914 further rate reductions were made. On June 22 the Board amended the rate schedule to provide for special but uniform rates for energy used solely between 11 p.m. and 5 a.m., such rates to be at a 30 percent reduction from those regularly established. On August 20 rates were reduced for lighting service; the charge for the first 20 kilowatt-hours was cut from nine cents to eight cents per kilowatt-hour, with corresponding reductions for increased consumption. Power rates were also reduced, with the cost of the first 100 kilowatt-hours being changed from five cents to 4.5 cents per kilowatt-hour and all over 10,000 kilowatt-hours reduced from two cents to 1.8 cents (this latter rate applied also to lighting service).

AGGRESSIVE SOLICITATION OF CUSTOMERS

The Board employed other devices to secure customers during the early years in the electric business. Secretary Geller, who managed the office for the Board, solicited customers when time permitted and received authorization in April, 1912 to employ another solicitor to assist in securing electrical business. This was apparently successful, as the Board authorized similar employment in succeeding years. In October, 1912 Geller requested permission to try out a plan of offering a bonus of $1.00 to the Board's electrical customers for every new customer they caused to be connected with the circuits under the jurisdiction of the city utility, the bonus to be in the form of a credit against their lighting account. This too was successful and the secretary was later authorized to continue the plan indefinitely. In 1915 the Board placed an advertisement in the *Eugene Daily Guard*, adding this means of promotion.

More direct attacks were made on the competition when it was felt that the Oregon Power Company was competing unfairly in the struggle to determine who would supply the citizens of Eugene with electricity. In a report to the Council on July 30, 1912 the Board wrote:

> Many of our citizens are to be commended for insisting on their rights, while some few prospective customers have been prevented from patronizing their own electrical plant, by reason of the fact that the corporation supplying gas in the city has threatened extortionate discrimination for furnished gas unless such prospective customers patronize this same corporation for electrical energy as well.
>
> It might be well for your honorable body to look into the gas franchise with reference to this method of endeavoring to force the public.

ACQUISITION SOUGHT

The Eugene Water Board also employed the state's recently enacted public utilities legislation to restrain the privately owned company. In January, 1913 it asked the Council to take action under the new law to secure the removal in a limited time of the poles, wire, lamps and other fixtures which had previously furnished the street lighting, but which were not now otherwise used. The Board contended that this unused equipment was a nuisance and hindrance to its construction, as well as being an unnecessary menace to life and property. In January, 1914 more serious action was taken when City Attorney G. F. Skipworth was asked to file a complaint with the Oregon Railroad Commission regarding discrimination as practiced by the Oregon Power Company in securing business in the city of Eugene, such practices not being in accordance with the company's rate schedules as filed with the Commission. In June of that year the Commission rendered a decision favorable to the Board.

While hearings before the Railroad Commission were in progress during the early part of 1914, the Board concluded that acquisition was better than competition and made overtures to the other utility. In a letter to its competitor on April 28, 1914 the Board stated reasons for purchasing the Oregon Power Company's electric distribution system:

> 1. One consolidated system is capable of more economical operation than two systems.
>
> 2. Such an agreement will bring about a more harmonious feeling between the citizens.
>
> 3. Eugene, by reason of its superb railroad facilities, can be made a manufacturing city provided proper inducements can be made, and we believe that a substantial reduction in power rates can be brought about providing the City has all the electric business in Eugene.
>
> 4. Each election, and especially the recent one, sustains the action of the citizens towards the principle of municipal ownership, and we believe that a dignified proposition can be made on a basis fair to yourselves as well as the City of Eugene.

The initial response of the Oregon Power Company was encouraging, although the private utility later declared it was not interested in selling. A letter dated May 18, 1914 stated that it was willing to open negotiations and that it felt a consolidation of the two systems desirable. The Board construed this as a disposition of Oregon Power Company to sell its distribution properties in Eugene and communicated its pleasure along with a request for an inventory of the properties. On July 29, however, the Oregon Power Company suggested that its earlier letter did not indicate a desire to sell, but that it was interested in negotiating with a view to acquiring the city properties.

WATER BOARD FORCES ISSUE

No immediate action was taken by the Board upon receipt of this notice, but it soon showed its willingness to force the issue. On August 20, 1914 it reduced electric rates and sent a letter to the president of the private firm declaring, "very evidently you do not understand the situation." It then suggested that the other utility arrange a meeting of its board of directors at a very early date and inform the Water Board of the decision. In conclusion, the Board stated ". . .we earnestly believe that a conference on any other grounds than the purchase of your distribution system would be futile and a waste of time."

The Oregon Power Company apparently realized the futility of further competition with the determined Water Board and in 1915 agreement was reached. Many proposals and counterproposals were made, with the Board speeding up progress with the declaration that it intended to install an auxiliary steam plant unless it could acquire Oregon Power Company's distribution system and an agreement to purchase standby power from the private firm. This added inducement led to an early conclusion of an agreement, with the Water Board signing a memorandum of agreement on December 13, 1914 to purchase the distribution system at a price of $150,000 ($50,000 down and $100,000 to be paid out of revenues—Oregon Power insisting that payment not come out of general taxes). A memorandum was also signed to provide for standby service, both contracts to become effective simultaneously. The Board sent the agreements to the Council for ratification and on February 11, 1916 a warrant was drawn for the initial payment and the competitive struggle was over.

A part of downtown Eugene looking toward city reservoir site and Lane County courthouse, c. 1912. (EWEB files)

WATER SYSTEM EXPERIENCES FINANCIAL PROBLEMS

It might have been expected that the practices brought about by competition and the acquisition of the duplicate electrical distribution system might have created financial problems for the Board, but such was not the case. There were financial problems, but they were in connection with the water department rather than the electric department. While the water system in its early days under the Eugene City Council had realized satisfactory revenues, the detailed financial analysis in the 1912 Annual Report demonstrated that revenue structures needed attention, as did the entire water system accounting process. At the time the Board took over the operation of the water utility, the operating expenses and interest up to that point were only slightly less than receipts from "water rentals," and no provision had been made by the Council for providing a sinking fund to redeem bonds when due. By the middle of 1913 the growth of revenues led the Board to conclude that its electrical operations were a decided financial success, a conclusion which subsequent reports reinforced. All electrical extensions were financed out of revenues, as was payment of current costs.

The Board insisted on keeping the water accounts separate from the electric, and set about doing something to remedy the financial situation in the older department. In 1912, during the latter part of November, the Board recommended to the Council:

> That the Council provide for removing (from) the Water Board the obligation of providing interest on $104,700, the amount of useless nonoperating property in the Water Department. (This had reference to a large area on or near Skinners Butte which had been included in the sale to the city of Eugene by the private water company.)
>
> If the Water Board is not relieved of this unnecessary investment, it will be necessary to have to resort to the rental of the butte property on long-term leases in order to raise the annual interest and sinking fund thereon.
>
> Since this amount is principally made up of the butte property, together with a large reservoir and some worn out machinery and a discarded well across the river, we would suggest that so much of the butte property as is not required by the Water Board for reservoirs, rights of way, etc. for the utilities in their care be turned over to the City Park Board for city park purposes, believing this to be in the better interest of the City than to lease this property out for other purposes.

This appeal to the Council was renewed from time to time and eventually the matter was referred to the people in the election of April, 1914. When the voters approved the proposal the Board was relieved of paying the interest and principal on some $93,000 of the original issue of bonds.

Although the water department was in financial difficulty in 1913, the Board was not deterred from moving ahead with improving and extending the water system. In the Annual Report for the period ending in December, 1912, recommendations were made to the Council that a $100,000 bond issue be placed on the April, 1913 ballot to make necessary improvements in the water system. At first the Council refused to do so, but when Board members took steps to get the measure before the people by use of the initiative process, the Council changed its mind. The citizens of Eugene, on April 7, 1913, authorized the Council to issue the $100,000 in general obligation bonds to improve and extend the water plant. The vote was 381 yes; 379 no.

Some difficulty was encountered in issuing and selling the water system improvement bonds. The Council had to be urged to issue them and when it did so, the bids proved unsatisfactory and were refused. The Board commended the Council on this action but urged it to offer them again with bids not to be received later than February 1, 1914. This time the bonds were successfully sold, although the purchaser had some afterthoughts, asking that the city levy a tax to cover interest and to establish a sinking fund for the issue. The Council refused to do so and the Board refused a later request to pass a resolution creating a sinking fund for the redemption of the bonds at maturity, although it indicated its desire to create a sinking fund for the redemption of all bonded indebtedness as soon as possible.

The funds from the bond issue were to be used primarily for building a new reservoir. In 1911 a joint bond issue had provided $27,000 for extension and improvement of water mains and for an additional pump, but the need now was for greater reservoir capacity and an extended distribution system as the city spread. Negotiations were begun in 1913 to purchase property on College Hill and in 1914 the arrangement was concluded and construction of the first College Hill Reservoir began.

There were in the early years some expectations that taxation would have to be resorted to to insure the success of the Water Board. The financial success of the electric department and the reduction of bonded indebtedness of the water department by turning over the non-operating Skinners Butte property to the Park Board eliminated the need of levying taxes to support that part of the utility serving domestic and commercial customers. The only taxes levied in connection with the operations of the Board, therefore, were those levied by the Council to pay for water and electricity for public purposes.

"ARM'S LENGTH" POLICY

Early in its existence, the Board had adopted a policy of "arm's length" dealings with the Council and various city departments. When the city installed street lights, it paid the Board out of tax funds when the Board presented the bill showing the expense of such installations. When the Park Board made an informal request for free water for drinking purposes at Riverview (now Skinners Butte) Park, the Board replied that it would be glad to furnish the water but that it would have to be paid for, although it would not object to carrying the account until the next tax receipts replenished the Park Board's funds. In 1915 the Board requested that the Council pay more for public use of water for fire protection, street flushing and other purposes, since on a cost basis it was securing water costing the Board $12,000, upon which the Council paid only $1,200. The Council refused to heed this request on the basis that it would have meant increased taxes. However, the Board held that the general public's gain was the water consumers' loss since "this unjust condition" meant that a reduction in water rates was not possible. The Board concluded, "From a business standpoint this condition is wrong, as each department, business or individual should bear its own burden of expenses. Public Utility Commissioners throughout the land recognize this principle and enforce this method as being a fair and just apportionment of expense. Our own Commission at Salem made a similar ruling at Salem over a year ago in the case of the city."

CONTAMINATION PROBLEMS OCCUR

There were problems with groups other than the Council and other city units. Private water supplies of the Oregon Power Company's gas plant, the Southern Pacific Railroad, and a citizen of Eugene were judged to be contaminated in 1912 and 1913, and the Board severed connections with the mains of these users until the situation had been remedied. One especially objectionable condition was created during the summer of 1913 when the discharge of the Oregon Power Company's gas plant into the Willamette River above the filter plant's intake resulted in the filtered water tasting like oil, gas and tar. Working directly and through the Council, the Board attempted to secure some abatement of this nuisance, but the gas plant manager refused to take remedial action. In September the Board was obliged to install a temporary intake pipe in another part of the river while also making a formal complaint to the State Board of Health.

The Board had some better neighbors whom it was pleased to serve. In July, 1915 the Board authorized sale of electricity to four or five people living on Camp Creek Road near the power plant. The construction of the lines was paid for by the new customers, but the Board later acquired them, paying in electricity. An important policy decision was made when the Board decided to charge these out-of-town customers at the old (and higher) rate schedule. In October of the same year the Board also agreed to extend commercial service to prospective customers in the town of Walterville when sufficient contracts were obtained.

During the early years of the Eugene Water Board the growth in average number of electric system customers was greater than for water system customers. However, it should be remembered that water was an essential household commodity long before the emergence of electricity. From 1912 until 1918 the average number of electric system customers rose from 492 to 2,593, an increase of 427%. Over the same period, average water system customers increased from 1,800 to 2,149, an increase of only 19%.

Employee relations were handled almost entirely by the Board during the early times. Wage rates for individuals were set by action of the governing body. It also adopted the resolution declaring that no one would be employed who used intoxicating liquors in any form whatsoever, either on or off duty, and that any employee found using intoxicants either on or off duty would be requested to hand in his resignation immediately. State legislation led the Board to apply the 1913 Eight-Hour

Labor Law to those engaged in physical labor, and in 1914 the Board brought all its employees engaged in hazardous occupations under the new Workmen's Compensation Law.

DYNAMIC YEARS

The years 1911 through 1916 were indeed dynamic. The Board established a philosophy of meeting problems with expedient action, a philosophy that has been carried forward to this day. On November 11, 1914 a new well was completed to serve as the city's water supply and pumpage from the river was discontinued. Undoubtedly the well was put in service in part because of the bad tasting water resulting from the gas company's discharge and it is also probable that increasing pollution on the upper reaches of the Willamette River was considered. The well was located on the north bank of the river across from the junction of Eighth and Franklin Streets.

The 1914 Annual Report included the following statement, "Municipal ownership is no longer a theory and a far-away dream, but it is a practical reality and has developed slowly and conservatively into a movement which is growing throughout the land with a force similar to other great economic movements of the present time."

In 1915, in addition to completing negotiations with the Oregon Power Company for the purchase of that system, the new College Hill Reservoir went into service in the latter part of September.

NOT WITHOUT CONTROVERSY

The activities of the Eugene Water Board were not without controversy. The purchase of the Oregon Power Company's distribution system was bitterly criticized by the *Eugene Daily Guard,* which contended that the price paid was too high. There also were numerous public discussions and controversies regarding electric rates. In response to this, on April 28, 1916, the Oregon Public Service Commission* was asked to make a survey of the Eugene Water Board and offer recommendations. The request was made by the Water Board, City Council, Chamber of Commerce, Fortnightly Club and other civic groups. The Public Service Commission agreed to undertake the survey and the work went forward throughout the remainder of the year.

What did those early years tell us about the community named Eugene? There were many dedicated persons who concerned themselves with not only the present but the future well-being of the community. Technology was state of the art if not beyond. The use of chlorine in water treatment to guard against epidemic disease was among the first in the nation, and as far as is known, the first in the Pacific Northwest. Much good work was done during those early times. Our present day municipal- utility operation can be very proud of its beginnings.

*Formerly the Oregon Railroad Commission

CHAPTER THREE

Coming of Age

DURING THE LATTER PART of 1916 and early in 1917, Eugene newspapers were filled with stories about the Great War raging in Europe. There were signs that United States neutrality could not be preserved and with the increasing loss of life of United States citizens to the German submarine war, it appeared inevitable that America would enter the conflict.

On March 20, 1917, the *Eugene Daily Guard* said editorially, "The supreme business of the hour is described in a single sentence: 'Get Ready!'," and on March 27 that paper printed a proclamation on its front page issued by Mayor W. A. Bell which said in part, ". . . and I further urge upon the young men in this community to respond to the call to the country and to fill up the ranks of the Navy and Army by volunteering in the service where possible." After the United States Congress voted overwhelmingly to declare war on Germany on April 5, 1917, President Woodrow Wilson proclaimed a state of war on April 6.

There were also frequent stories in those times about the Mexican Revolution and the activities of Pancho Villa. Also in the minds of newspaper editors was prohibition and it appears that the *Eugene Guard* favored this constitutional amendment.

As the Eugene Water Board matured, farsighted and sound policies were adopted, assuring success for the muncipal-utility organization. In May, 1917 the Oregon Public Service Commission issued a report on the operations of the utility, making recommendations as to changes in the operations of the general office, water department and electric department, as well as recommending changes in rates. When C. W. Geller resigned as general superintendent-secretary the following April, the governing board declared:

> The Water Board is deeply appreciative of the fact that during the years of Mr. Geller's service the water and light utilities have grown from a feeble beginning in the face of stern opposition to strong, dependable maturity and now occupy the field without competition and without measurable opposition.
>
> The Water Board is also ready to affirm its belief that the satisfactory results obtained have in a very considerable degree been due to Mr. Geller's untiring efforts, integrity, and unswerving loyalty.

E. U. Lee resigned from the Board is September, 1919. He was the sole remaining member of the body appointed by the Council in 1911. In accepting his resignation, the other Board members resolved: That his efforts have been untiring in putting the Water and Electric Departments on a sound business basis, and he has held steadfastly to a strict business policy even at times in the face of severe adverse criticism.

Carl A. McClain was appointed general superintendent-secretary on April 22, 1918. McClain attended and graduated from the engineering school at the University of Oregon and later received a master's degree at Cornell University. Having worked for the Eugene utilities in various capacities on a part-time basis between 1914 and 1917 while a University of Oregon faculty member, he was a logical choice for the position. During McClain's tenure the Walterville generating plant was enlarged, new generating facilities constructed near Leaburg and a steam plant built to provide standby service for the electric utility. New water supply systems were built along with a number of new reservoirs and a 12 million gallon per day filtration plant went into operation on the date of McClain's death, which occurred August 5, 1932.

ADDITIONAL GENERATION NEEDED

Growth in the number of electric system customers prompted the Board to consider increasing generating capacity. Conferences were held with a consulting engineer in 1921 and in 1922, the Board met informally with the City Council to discuss the question of building a steam plant as an alternative to entering into an agreement with another utility for the purchase of energy above the Walterville plant's ability to produce. This issue was resolved for the time being by the approval of a standby agreement effective January 1, 1923 with the Mountain States Power Company (successor to the North Idaho and Montana Power Company, which in turn had acquired the Oregon Power Company). Although standby power could be secured at favorable rates, it was not a desirable long-range solution to the expanded demand for electric energy and the Board took other measures. At the time the standby agreement went into effect, purchase of additional generating equipment was authorized for installation at Walterville and was in operation by March, 1924. Recognizing that this was not a permanent sol-

Carl A. McClain, general superintendent-secretary of the Eugene Water Board from April 22, 1918 until his death on August 5, 1932. (Kennell-Ellis photo)

ution, the Board on April 30, 1924 authorized the investigation of sites for the development of additional power.

The search for potential power sites begun in 1924 eventually led to the construction of new generating facilities near Leaburg. The first site seriously considered was property near Bear Creek where an "Upper Marten Rapids" power plant was projected. Core drilling and other investigations conducted during 1926 revealed many problems at this location. In March, 1927 exploratory drillings were conducted near Leaburg and the results were such that the Board was encouraged to submit the question of building power-generating facilities at this site through the City Council to the people of Eugene in a special election held June 28, 1927. Eugene voters approved the plan along with the bond issue of $1.25 million needed to finance the construction. Contracts were let early in 1928 and on January 6, 1930 the Leaburg Power Plant went into service. The so-called Rennie property near Bear Creek was not forgotten and the Board acted to purchase it in April, 1933 as a possible future power site.

Shortly after the Leaburg plant went into operation, the Board moved toward the construction of its own standby facilities. Standby contracts had been in effect with Mountain States Power Company up until this time, but the Water Board felt that the private company's rates were excessive and in April, 1930 the designing of a steam standby plant was authorized with contracts let early in July of that year. In late summer, 1931 the steam electric generating plant went into service and the electric utility was in a position to provide its own energy at all times at a lower cost than that available from other sources.

CLEAR LAKE WATER SOURCE DISCUSSED

Increasing demands for water during the 1920s led to a search for additional sources. In 1924 the Board discussed with the City Council the prospect of securing more water through a "Clear Lake Water Supply Project." Nothing came of this nor of a proposal to secure well water in the territory north of Eugene between the Willamette and McKenzie Rivers. In January, 1925 the Board made a decision to proceed with the McKenzie River Water Supply Project, the water to be taken from the McKenzie in the vicinity of Hayden Bridge. A special election was held in April and voters authorized the proposed extension of the water system and the issuance of $375,000 of "McKenzie Water Bonds" for the purpose. The contracts were let in September and during the summer of 1927 the project was completed.

Increased water use required new reservoirs and other additions to the water system. The average number of water system customers in 1924 was over 3,500 and water consumption per customer had been increasing over the years. In April, 1924 the Fairmount Heights Reservoir was approved by the Board, although the contract was not let until August, 1925. On November 24, 1925 a contract was let for a larger reservoir on Skinners Butte. Both projects were financed out of a $135,000 issue of "Water Extension Bonds" which were authorized by the voters on April 15, 1925 as a companion measure to the McKenzie River Water Supply Project. To provide water service to the Southern Pacific shops and terminal which had come to Eugene in 1926, the Board asked for an additional authorization for $30,000 of bonds to be known as "Water Extension Bonds, 1926." After approval was secured at an election held July 6, 1926, the Board went ahead with the construction of water transmission and distribution lines. Money received from the 1925 bond issue was employed for similar purposes elsewhere in addition to the construction of reservoirs.

To maintain the quality of the water being used in ever- larger quantities, it was also necessary to add filtration capacity during McClain's administration. In 1924 the Board acquired additional property along the Willamette River in Eugene to provide room for settling tanks (as well as for the proposed steam-auxiliary plant). In September, 1931, just after the steam plant had gone into operation, the Water Board voted to proceed with McClain's plan for a new filtration plant to be built without additional bond issues. Going into service on August 5, 1932, the plant was named the Carl Arthur McClain Filtration Plant in honor of the man who died that day after suffering a stroke while on duty at the Eugene Water Board office.

Nowhere in the files or news accounts of Eugene Water Board activities is there found remonstrance against the concept of bringing water to Eugene all the way from the McKenzie River, a distance of over seven miles. Undoubtedly, there was more upstream pollution on the Willamette River than on the McKenzie. It is also true, because of the subterranean storage of water in the Clear Lake vicinity, that the stream flow of the McKenzie tends to be much more stable throughout the year than is the case with the Willamette. Interestingly enough, however, the Willamette in recent times has been declared a satisfactory source of water for domestic use.

COMMODITY USE PROMOTED

The increased demands for water and electric energy during these years were more than the product of the growth of Eugene. Also involved were measures adopted by the Board designed to increase the use of these services. General electric-rate reductions and special rates for various types of electric service were utilized. Electric ranges were sold by the Water Board and cooperative-selling campaigns were embarked upon in conjunction with local merchants. Both water and electric service were extended beyond the city limits, in part to gain greater efficiency in the electric generating system and the filtration plant.

The 1917 report of the Public Service Commission recommended a reduction in electric rates at that time which was adopted by the Board and was followed by a succession of additional decreases. In 1920 the first increase in electric rates was made—from 75 cents to 90 cents for the first six kilowatt-hours—when it appeared to the Board and to the Public Service Commission that the existing rates would create a deficit.

Beginning in 1925, however, rates were reduced a number of times. On January 1, 1925 the pre-1920 rates were restored and on March 1, 1926 electric rates reached an all-time low. Further reductions were made in 1931 amounting to about 16 percent.

The trend in water rates was also downward between 1917 and 1932. There was an increase in 1920 and when the 1925 bonds were sold, rates were increased for sinking fund purposes. However, there were reductions in 1929 and 1932, and by 1932, water rates also were at an all-time low. Also in 1932, because of the Depression, the city of Eugene was having difficulty meeting its many bond obligations for strictly municipal expenditures. As a cooperative measure, the Water Board voted to give the city government a 25 percent discount on water and electricity used for municipal purposes.

Certain special rates served to promote the expanded use of electrical energy. In addition to the existing special rate for "off-peak" power during the night hours, favorable rates were extended to bakeries in 1918 to apply between 10 p.m. and 6 a.m. as an encouragement to the installation and use of electric ovens. In July, 1919 a flat rate of $3.50 per month per kilowatt of demand on electric water heaters was established when such water heaters were used in conjunction with electric ranges. The water heaters were required to be interconnected with the ranges so that only one appliance could be used at a time. This measure increased the use of both electric ranges and water heaters, while at the same time tending to control system peak use of electricity.

In 1917 the Board had entered the electrical appliance business. On July 9 of that year a motion was adopted in which then General Superintendent Geller was instructed to carry a stock of electric ranges and other appliances for sale, either on a cash or deferred payment plan, in order to increase the electric load for cooking purposes. Although this policy was reaffirmed in 1919, a resolution was approved 10 years later, in 1929, which discontinued the sale of electric ranges because local dealers assured the Board that they could provide satisfactory sales and service to those customers purchasing these appliances.

During the early 1930s, however, additional promotional plans were adopted, including cooperative campaigns with local electrical dealers. The National Electric Light Association had outlined a special electric-refrigerator campaign for adoption by utilities in cooperation with dealers and the Board authorized participation in a Eugene campaign during March, 1930. In September of that year McClain was given discretion to employ a special demonstrator to assist in encouraging use of electric ranges and other current-consuming devices. To further promote the use of electric ranges and the consumption of electric energy, a plan was approved in November, 1931 for furnishing free wiring for electric-range service. Promotion of another kind had occurred during the summer of 1931 when the Board contributed $200 to a development fund dedicated

September 5, 1926; 30-inch line to bring water from McKenzie River to the Eugene system. Looking east to Boller Corner, engineers noted crook in pipe "supposed to be laid straight." (EWEB files)

September 3, 1925; 30-inch line at Willamette River crossing. Pipe is in position for lowering. (EWEB files)

to locating the National Veterans Home in the vicinity of Eugene. The Board apparently assumed that its contribution and more would be returned in utility revenues if the campaign were successful.

After the first extensions of electric lines to consumers in the Walterville area, the Board continued to make energy available to those outside the Eugene city limits. By 1920 the policy for electrical line extensions was well in place. Petitioners could acquire service by entering into a contract guaranteeing that one-third of the cost of extension would be forthcoming in revenue each year for a period of three years. Somewhat later the figure was increased to 40 percent of the cost of the extension in each of three years of service delivery, but again changed when the Board adopted the Willamette Extension Rule published by the State Public Service Commission in December, 1925. However, in March, 1927 the Board adopted its own formal rules governing the extension of electric lines into rural territory. Plan A (as the rules were called) provided that the Board would finance and construct at its own expense any extension requiring an investment of not more than $60 per customer or assurance of two and one-half times estimated annual revenue if such revenue was at least $24. Plan A was replaced by Plan B in June, 1930. The new rules provided free extension when the cost of such extension did not exceed $75 per customer or not more than five times estimated annual revenues, the Board reserving the right to request a contract assuring that the customer take and pay for service during a period of five years. In both Plans A and B, customers were required to pay all amounts above the free limits in advance of construction.

In 1917, when the Public Service Commission made its report to the Board, it found no financial problems although it recommended some changes in financial policies and practices. Rather than establish a sinking fund for the retirement of bonded indebtedness when it came due, the Board had chosen to retire the debt by purchasing the bonds before maturity. Since surplus funds were kept in local banks which paid two percent interest on part of the balance, the Commission suggested that these funds might be better employed in the purchase of bonds which could, if necessary, be used to secure bank loans. When it purchased its own outstanding obligations, the Board was obliged to retire them, which might have resulted in a cash shortage in an emergency. The recommendation was not immediately adopted, but beginning in April, 1918 a policy was adopted of purchasing Liberty Bonds for sinking fund purposes in order to pay off the bonds coming due in 1920 and 1926. In 1921 the Water Board also began to invest in other acceptable securities including obligations of Eugene and Lane County. Sufficient funds were accumulated as a consequence to enable the bonds to be retired at maturity without financial strain.

Some problems were encountered with the sale of bonds to finance construction projects authorized during the 1920s. The McKenzie water

August, 1923; photo shows break in Walterville Plant penstock No. 2. Water Board employees George Easton and Edward Howells were at site when rupture occurred. (EWEB files)

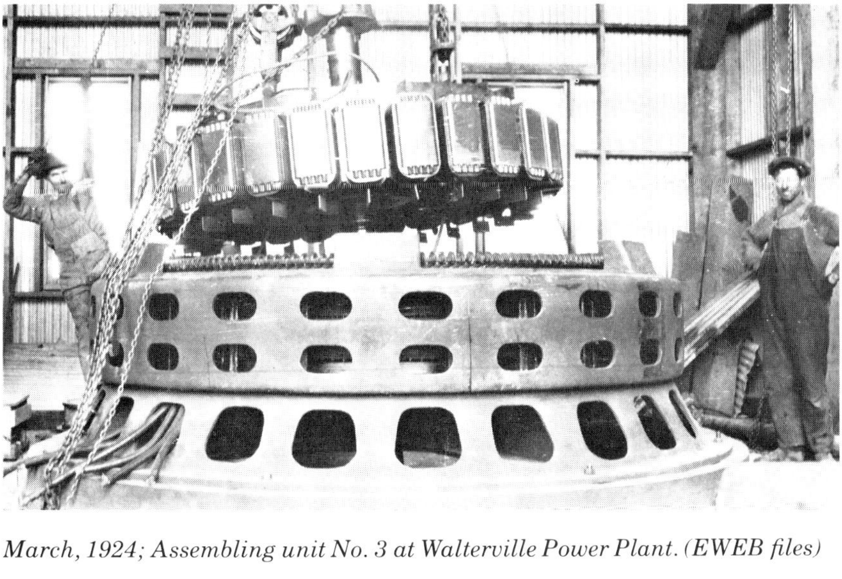

March, 1924; Assembling unit No. 3 at Walterville Power Plant. (EWEB files)

August 23, 1928; Walterville Power Plant—view from bridge over tailrace. The plant was no longer called "Matlock Station." (EWEB files)

bonds and the water improvement bonds were sold in 1925 without difficulty, but the power and light bonds authorized in 1927 met with trouble. The problem lay not in the quality of the Water Board bonds but in the rising interest rates in the bond market which made it impossible to sell the authorized bonds at par. The construction of the Leaburg project was not seriously hampered thereby, since there was over $500,000 in the surplus accounts of the electric department. Cash and investments were on hand to begin construction and as necessary, funds were borrowed from the water department at three percent interest. After the Board experienced difficulty in selling the first block of authorized power and light bonds (about March 11, 1929), it sought and obtained authority by the end of May to sell interim power and light bonds at higher interest rates, and on June 3, $500,000 of such securities were sold. In September, 1929 another $200,000 of interim bonds were sold, another $200,000 in October of that year and a final block of $300,000 in August, 1930, resulting in a total sale of $1.2 million of interim power and light securities.

The interim power and light bonds sold during 1929 and 1930 were of short maturity—five and seven years—and consequently posed a fairly immediate financial problem. The Board was able, during 1931, to refinance $600,000 of interim power and light bonds, bearing interest rates of 5½ percent and 5¾ percent, with $600,000 of longer-term regular power and light bonds, bearing interest rates of 4½ percent and five percent. Also, revenues accumulated from sales of electricity were substantial enough to ward off any threat of insolvency.

THE BOARD IMPOSES TAXES

Due to the sound financial situation of the water and electric utilities tax levies to pay bonded indebtedness were not necessary, although throughout the entire period of 1916-1932 the Board levied taxes for public use of electricity and water. The mayor of Eugene informed the Water Board in November, 1916 that the Common Council would not provide anything in its levy to pay for the public use of light and power as it had been doing in the past and that if the Board desired to be paid for this serivce, it would have to make the levy itself. Thereupon the Board, acting as the commissioners of the Department of Water and Light of the City of Eugene, made an estimate of the amount which would have to be raised by taxation for the purpose of defraying all authorized demands on the city treasury for the public use of water and electricity. This continued until 1932 and in each year there was some increase in the estimate as the public consumption of the services of the utility increased. On October 11, 1932 City Attorney S. M. Calkins informed the Board, after it had requested an opinion, that the Board was not a tax-levying body and that it did not come under the provisions of the local budget law, that it was not required to make a report as a department of the city in the city's budget, as it was self-supporting, and that the levy was to be made by the city proper in its budget for services rendered the city.

During the years in which the Eugene Water Board was regarded as a tax-levying body, it called meetings of taxpayers to consider its proposed budget for the coming year. Either confidence in the Board and a laymen's budget committee or lack of interest resulted in all taxpayers' meetings but one being attended by no taxpayers other than the members of the budget committee. The sole exception was the meeting held in November, 1931 when a group complained of the maintenance of employees' salaries and wages at previous levels during a depressed period when other workers were either out of work or receiving reduced incomes. No change was made in the budget at this time but the Board and its employees were later to take measures to meet this criticism.

There were community relations problems between 1916 and 1932, and the Board yielded to the requests made by various groups whenever necessary or desirable. Local business and labor groups required mollification. The Chamber of Commerce and other community organizations were influential in the determination of the policies of the Board. The Common Council, the University of Oregon and other government groups required increased attention.

When the Water Board had work to be performed it was aware of the importance of giving local workers a fair share of employment. In 1928 McClain was authorized to call the attention of the contractor on the Leaburg project to the decrease in percentage of men from Eugene and vicinity who were employed on that job and to offer to assist the contrac-

November 27, 1929; Leaburg dam and canal intake under construction. (EWEB files)

July 5, 1931; Leaburg power house. (EWEB files)

tor in seeking and hiring local help. This apparently did not have the desired result and the Board—after meeting with a committee representing the carpenters' union—authorized the investigation of alleged practices of employing other than local labor and paying wages below local scale. When the new filtration plant was authorized in September, 1931, its construction was justified in part because it would provide employment which was much needed in the community. Although the Board declined a later petition requesting that the job be given to a local contractor, it gave the job to a Portland company only after the firm agreed not to employ at any one time more than four men who were not residents of Eugene and vicinity. In December, 1931, after representatives of local workers had complained at a taxpayers' meeting of the relatively high wages received by Board employees during the depression, McClain was authorized to proceed with plans for unemployment relief work with the common labor hired to be paid from relief funds contributed by Water Board employees.

November 20, 1929; Hugh P. Currin at Leaburg project site. Currin is only person for which an EWEB facility is currently named—the Hugh P. Currin Substation. (EWEB files)

"FEAR NOT"

The success of the resulting employee effort is not known. However, one Board employee demonstrated concern for citizens of the community who were experiencing difficulty during the depression. Gordon B. Fish, who had gone to work for the Water Board in June, 1917 and had served many years as an electric serviceman, had the task of terminating customers for non-payment. On several occasions, rather than terminating service, Fish paid the utility bill himself. As late as 1957, he was being repaid by some of those who had benefited from his benevolence years earlier.

Gordon Fish had earned the nickname "Fear Not" among his fellow employees. This derived from the legend that Fish, in order to determine if a wire was hot, would touch it! He was a familiar sight in the community as he made his service calls on a motorcycle equipped with sidecar for tool storage. His nephew, Eveard Fish, was also a long-time Board employee.

Interests of local businessmen were given attention by the Board. Not only did the Water Board discontinue competing with local appliance dealers in the sale of electrical appliances, it also adopted in 1932 the practice of giving the contract for electric lamps—which had earlier been purchased directly from General Electric or Westinghouse—in rotation to Eugene electrical contracting firms exclusively selling electrical supplies including ranges. When four operators' residences were to be constructed at the new Leaburg plant, Eugene contractors were utilized for various parts of the work even though their bids were higher than those of certain Springfield and Salem contractors.

RELATIONS WITH CHAMBER IMPROVE

During the 1920s there was an improvement in relations between the municipal utility operation and the Eugene Chamber of Commerce. The businessmen's organization, known in the early days as the Eugene Commercial Club, had opposed the establishment of a municipal utility in 1905. After formation of the Water Board in 1911, this hostility continued throughout the first decade of the Board's existence. In 1914 the Commercial Club launched a campaign to bring more factories to Eugene and urged the Water Board to offer power to prospective factory establishments at no cost or sharply reduced rates. The Board insisted that providing free power would bankrupt the utility and incurred considerable criticism when it rejected the request. In 1916 the business group, which had been renamed the Eugene Chamber of Commerce, adopted a resolution to the City Council requesting an order that the chamber be given the power to examine the books of the Water Board and issue a public accounting on the efficiency of the utility operation. The Board did not "appreciate the insinuation" that it was necessary to

May 29, 1931; erection of large doors at east end of standby steam plant. (EWEB files)

January 2, 1932; filtration plant mixing chambers ready to pour. (EWEB files)

secure an order to examine the books since they were open to the public "at any and all reasonable times." The dispute between the Board and the Chamber of Commerce about the proper level of water and electric rates led to a request directed to the Oregon Public Service Commission for an investigation and recommendations. The request was submitted in April, 1916 and as pointed out early in this chapter, the report of the Public Service Commission was considered and duly acted upon by the Board.

Notwithstanding the earlier difficulties with the Eugene business community, the Board in 1926 pledged $10 a month to the Chamber's development fund "in view of the value to the community of the work being carried on by the Chamber of Commerce . . ." The practice of subscribing to the development fund of the Chamber was continued in succeeding years.

SOME FREE SERVICE GIVEN

Although unwilling to give large and continuing donations of water and electricity, the Board, up until the end of 1932, did on occasion provide free water and electric service. In May, 1917 the general superintendent-secretary was authorized "to furnish light and water service for the two patriotic organizations, one known as the Girls National Honor Guard, Eugene Branch and the Eugene Branch of the Red Cross Society, on a free basis." Later donations of services were made to various organizations conducting affairs for charitable purposes. An outdoor theater operated by the University of Oregon received electric current without charge in the summer of 1918 "in consideration of the valuable service rendered the Board by the University." During 1924 water bills were forgiven during July and August for customers awarded lawn prizes by the city beautification committee of the Chamber of Commerce Auxiliary. In the early 1930s, however, the Board began to deny requests for free service and reductions in rates for charities and benefits. Finally, on November 14, 1932 the Board refused the city firemen free lights for their proposed charity ball and adopted a resolution that "all like requests that might come up in the future be refused for the reason that the Board cannot see its way clear to grant all requests of this nature and that it would not be fair to make an exception."

Relations between the Eugene Water Board and the Common Council were generally cooperative although sometimes strained because of the financial question. The Council had joined in the request to the Public Service Commission in 1916 for a study of the Water Board and it was later stated by the state agency that the city was not paying enough for public uses of water and electricity, a matter which had been in contention. Thereupon the Council put the burden of levying the tax for this purpose on the Board. The Board began levying taxes and continued the practice until 1932, when the city attorney declared this to be properly

August 2, 1932; interior of new filtration plant—picture taken three days before death of Carl A. McClain, the man for whom the plant was named. (EWEB files)

August 27, 1932; view from north end of Ferry Street—showing both filtration and standby steam plants, a fitting memorial to Carl McClain's tenure as general superintendent-secretary. (EWEB files)

the province of the Common Council. Arrangements made in 1916 also reflected the policy of "arm's-length dealing" which the Board had adopted in its relations with the Council. The Board agreed to rent space in the new city hall quarters, located in the old Eugene High School building at Eleventh and Willamette which had been recently acquired by the city, with the understanding that the Council would pay for water and electric service for the entire building.

In the general sense, however, relations between the two bodies were friendly and cooperative. The Board and Council worked together in the early 1920s on improving street lighting. In 1926 the Water Board constructed a water main along the Pacific Highway north of the city limits when the city was threatened by litigation on account of pollution of wells from discharge of the city's sewers. This was in contradiction to the established water main extension policy, but the Board stated ". . . the Eugene Water Board, being a department of the city of Eugene, is obligated to serve the broader interests of the city as a whole, even when said interests involve the setting aside of fixed policies in specific instances." On October 28, 1929 a resolution was adopted granting a rebate not in excess of 25 percent to any municipality using light and water products of the Board in excess of $20,000. As the city of Eugene was the only such customer, this was apparently more a matter of easing the financial load on the city than encouraging greater consumption. This policy went into effect in 1930 and continued for a number of years. The City Council, as well as the Board, occasionally made concessions. In 1918 the Council permitted an increase in the flat rate charged the city for fire protection in order to enable a sinking fund to be built up to retire the water bonds on the Skinners Butte property taken over by the city without the Water Board levying a tax for this purpose. From the standpoint of the citizens of the community, the fiscal arrangements between the Board and the Council probably were not of crucial importance. The citizens, either as utility ratepayers or as taxpayers, paid for the services rendered by the two bodies.

Demands were made on the Board by school authorities. Both the Eugene Public Schools and the University of Oregon had asked for combined meter readings on the properties under their control. The existing method involved separate meters as well as bills for each building. Apparently the rationale of the school officials was that they would be able to incur a single minimum charge rather than the more costly multiple charges. In December, 1928 the Board granted this oft-repeated request and adjusted the bills back to January, 1924 using the combined basis. The Board in October, 1928 politely declined a request from a campaign committee attempting to raise a large sum in gifts for the university, stating that it could not consistently make either a cash contribution or a concession in rates. However, in 1932 it made substantial contributions of cash, up to $25,000, and the time of its employees to campaign against the infamous Zorn-MacPherson initiative which would have consolidated the University of Oregon with Oregon State College on the latter's campus, thereby depriving Eugene and the Water Board of a large segment of the community including a substantial number of customers of the Board.

The Zorn-MacPherson initiative was one of the most dramatic episodes in the history of Eugene. In addition to moving the University of Oregon to Corvallis, the initiative would have eliminated altogether the normal school at Monmouth and restructured the schools at La Grande and Ashland as "junior colleges." The plan also called for removing legal education from the State System of Higher Education and consolidating the University of Oregon Law School with Willamette University in Salem.

Primary sponsor of the measure was the Marion County Tax Equalization League, of which Henry Zorn of Aurora was president. The League was supported by Hector MacPherson, member of the Oregon House of Representatives from Linn County and former Oregon State College economics professor.

Notable among Eugeneans opposing the measure were University of Oregon Law School Dean Wayne L. Morse and *Eugene Register-Guard* Managing Editor William M. Tugman. Save for the *Corvallis Gazette-Times*, the press in the state was uniformly opposed to the measure as was Governor Julius Meier and at the general election in November, 1932, the measure was defeated six to one.

While the Eugene Water Board minutes reveal that up to $25,000 was authorized to help defeat Zorn-MacPherson, it is not known how much was actually spent. Perhaps the sharp reaction of the community as a whole and the ensuing momentum that carried through the summer of 1932 obviated the need for at least part of the authorized amount.

SUPPORT GIVEN BOARD

Good relations which the Water Board generally enjoyed with the public were demonstrated by an episode occurring in 1930. In June of that year E. J. Adams, formerly of Eugene and at that time an attorney for the Federal Trade Commission, proposed a plan to "industrialize" the Eugene area—with the proviso that the electric system be sold to eastern financiers. The Board immediately rejected the plan, publishing a statement containing its reasons in the local newspapers on July 1, 1930. Both Eugene newspapers took vigorous stands against the move, pointing out that any industries wishing to locate in Eugene already could have the advantage of low rates under municipal ownership and that sale of the electric utility would be no assurance that successful industries would be established. The lack of any significant local support for the plan was instrumental in its failure.

During the years from 1917-1932 there were many changes in the

Board's relations with its employees. Pay rates and methods of payment were improved. Insurance coverage, other benefits and recognition were provided. The improvement in wage and salary administration commenced in 1917 when a schedule of wages for employees in the water and light departments and a scale of monthly salaries for office workers was adopted. Prior to this time all wages and salaries were matters of individual negotiation and decision by the Board. In September, 1918 the Board adopted a policy of adjusting salaries every three months on the basis of a cost-of-living index computed from prices charged for various items by local merchants—a pioneering measure in the area. Exceptions were made for the line crew, which was to be paid the prevailing union rate for cities the size of Eugene—also a new wage policy. Wages were raised occasionally during the 1920s, and when the depressed conditions of the early 1930s justified the move, a general salary reduction was made effective May 1, 1932 throughout the organization.

Insurance coverage for employees was provided under both the state of Oregon and private contracts. Employees engaged in hazardous duties had been covered under insurance provided by the State Industrial Accident Commission previously, and in March, 1929 the Board extended this coverage to all of its employees after investigating the reaction of workers to a plan of group disability insurance presented by a private company. The hazardous nature of the work performed by many employees had earlier—in June, 1926—influenced the Board in adopting a motion approving the proposal of a private insurance company providing voluntary group-life insurance for employees. However, the Board was not yet ready to contribute funds to greater insurance coverage for its employees. In March, 1932 the Board "laid on the table" a petition from a number of its employees requesting medical, surgical and hospital insurance. The governing body declared it impractical to furnish such benefit because of the small number of employees to be covered. The Board also stated that contracts would not be considered from companies which conducted personal solicitations of petition signatures among employees. The Board also declared that if a contract were to be accepted, it would be selected on the basis of the best interests of the employees and the utility operation.

Previously established services and benefits to employees were continued and new ones inaugurated. Additional housing for employees at both the Walterville and the new Leaburg plants was provided. Employees on the Water Board payroll "joining the colors" during World War I had their positions held open for them. Employees were granted a Christmas gift of $10 beginning with that holiday in 1922.

McCLAIN PRAISED

The Board made some attempt at providing an orderly succession of management, but time ran out. On January 11, 1932 McClain was given a four-year contract terminable only by his death or mental disability. This contract was made in recognition of the fact that McClain had guided the great growth of the organization during his tenure and that no one else could safely take over the management without a reasonable period of training and that a four-year contract would provide such a term. Unfortunately, McClain died less than eight months later. On the day of his death, August 5, 1932, his presumptive successor, J. W. McArthur, was appointed acting general superintendent-secretary, and on October 10 of that year he was named general superintendent-secretary.

Following McClain's death the Board adopted the following resolution:

> In view of the untimely death of Carl Arthur McClain, who for many years has served the Eugene Water Board as General Superintendent-Secretary, and who for that period of time gave special thought to the problem of supplying the City of Eugene with pure and wholesome water, and whose life work seems to have been culminated in the practical completion just before his death of a new and modern water filtration plant for Eugene, it is moved that said new filtration plant be and is hereby named the Carl Arthur McClain Filtration Plant.

The Board also passed resolutions in memory of Mr. McClain praising his judgments, kindness, friendliness and sterling character and ability.

CHAPTER FOUR
Assuming A Leadership Role

THE PAGES OF THE *EUGENE REGISTER-GUARD* during the early months of 1932 reflected the mood of the times. There were many stories about the Great Depression, the Sino-Japanese War and growing national interest in the repeal of prohibition. On February 29, 1932 a *Register-Guard* editorial warned that even with the return of prosperity, unemployment would remain at a high-level because of the mechanization of industry.

The Eugene Water Board had its turn at a front-page banner on April 26 when the following headline appeared:

Weather: Showers
Eugene Register-Guard
Home Edition
M'ARTHUR RILED BY RAP AT POWER
COMPENSATION RATE INCREASE DISLIKED HERE
Massie Conviction Demanded By Sta
ANIPULATION OF STOCKS BY POOLS RAPPED
HONOR TRIAL JURY LEAVES COURTROOM
SAYS RATES HERE STAND COMPARISON

(University of Oregon Library)

The accompanying story read in part:

> In a talk before the Eugene Rotary Club Tuesday on business and politics and their relation to the depression, H.H. Manny of Seattle

took a potshot at Eugene's municipal power rates and drew a prompt but polite challenge from J. W. McArthur, engineer of the Water Board.

Manny's remarks centered on blunders of municipal ownership in Seattle, but he pointed to expensive mistakes in other cities.

"You people here, I understand, are much in the same situation. You have a municipal power plant and your rates for power and light are higher than those of surrounding towns on the coast."

McArthur asked for the floor when Manny sat down.

"The points of the speaker are well taken," said McArthur, "with one exception which I cannot permit to go unchallenged. I think he has been misinformed when he says that our rates are higher than those of surrounding towns.

"Our rates in Eugene are not high; they are much lower than the average. We would be glad to have Mr. Manny or any member of this club go over our records and make comparisons, because the facts do not bear out his conclusion."

Thus the Eugene Water Board was attracting the attention of the press and the citizens and was beginning to take its place as a well-known utility operation in the Pacific Northwest.

On August 7, 1932 the *Register-Guard* eulogized Carl A. McClain in an editorial, a reproduction of which follows:

CARL A. M'CLAIN.

THE efficiency of Eugene's municipally owned utilities was in a large measure due to the quiet and effective personality of one man, Carl A. McClain. Mr. McClain was one of those very rare men who can take a political position and lift it above politics. The death of such a man is a great loss to the community.

The success or failure of any public enterprise depends greatly on the men in it. Whether any particular department of government is good or bad depends mainly on whether people have formed good or bad habits concerning it. If an office is held by a man whose personal ambitions outrun his zeal for the job at hand, confusion results.

Carl McClain took the management of the Eugene water plants and power plants because he was an engineer and because he regarded the administration of those plants as primarilly an engineering problem. He was of that finest type of engineers—a man who enjoyed his job not for the money he could get out of it but for the satisfaction of seeing it well done.

The splendid plants which Eugene has erected out of municipal earnings is a monument to Carl

McClain's quiet genius and vision. He could see years in advance what would have to be done and he could plan how it should be done and he could do it. Gentle and modest, McClain had a gift for friendships. He had also a certain firmness and soundness in his ideals which commanded the respect of those who knew him.

Carl McClain had an important part in building one of the finest traditions of Eugene, the tradition that efficient management of public business is NOT impossible.

(University of Oregon Library)

Joseph W. McArthur, McClain's successor, was a graduate of the University of Oregon engineering school who had had a long association with the Eugene Water Board prior to becoming general superintendent-secretary. He was employed as a part-time engineer during 1908 when the Walterville power plant construction was begun. In April, 1913 he became a member of the Board, the first person to be elected by the people. Ten years later, on April 9, 1923, he resigned his Board membership to become chief engineer, a position he held until appointed to succeed McClain following the latter's death in August, 1932. Thus qualified by training, long experience and deep interest, he was well-suited to deal with the problems of the Depression decade and the war period immediately following.

Throughout the troubled 1930s McArthur was aided by a strong Board. Three men—L. D. Pierce, Percy W. Brown and Orlando J. Hollis—served as members from before McArthur's appointment until 1941 or after. Hollis, a professor at the University of Oregon Law School, played an especially strong role and there were few motions adopted during his tenure that he did not propose or second. The combination of a vigorous governing body and well-qualified chief administrator aided the maintenance of a healthy and growing organization during a period when other businesses were experiencing great difficulty.

FEDERAL ASSISTANCE SOUGHT

The completion of the Leaburg hydroelectric plant in 1930 and the steam electric plant in 1931 relieved the new general superintendent-secretary and the Board of the need to construct additional electric generating capacity. However, the availability of federal funds led to the consideration of construction. Early in 1934 the Board applied to the Federal Emergency Administration of the Public Works Administration for a 100 percent grant of federal funds to begin the Rennie Dam power project on which rights had been filed in 1932. It was proposed that the PWA build the plant—at an estimated cost of $12 million—and that it be leased to the Water Board with an option to buy. This application was denied and no further action was taken on the Rennie site plan. Early in 1938 the Board discussed whether to purchase power from the recently established Bonneville Power Administration, which shortly would have inexpensive power available, or to build more steam generating facilities. In 1940 it reached the conclusion that it should do both and the second unit of the steam plant was completed in the fall of 1941. No other plants for the generation of electricity were constructed until well after World War II, although improvements were made in the transmission and distribution systems.

Neither was there as much construction activity in the water department as there had been in the previous period. The increased use of water by 1941 prompted the Board to use booster pumps on the McKenzie River water supply line and to begin studies of the feasibility of securing water in the future from the upper McKenzie. Studies concluded in 1944 demonstrated that the cost of installing a pipeline to Olallie Creek (a possible source of supply some 66 miles by car from Eugene) would not be less than $13.6 million. Preparations were made for filing rights to this water and also that from Anderson Creek, located above Olallie Creek, but the immediate solution was to install additional pumping equipment at Hayden Bridge to speed up the flow in the 30-inch supply line. Other construction taking place in the water system prior to 1946 was also in the transmission and distribution systems. The exception was the second College Hill Reservoir. On October 1, 1938 the Board accepted an offer of the PWA (which the Board had sought) to provide a grant of 45 percent of the cost of constructing a reservoir of approximately 15 million gallons with an additional 100,000-gallon tank for higher elevation service.

BOARD ENTERS STEAM BUSINESS

In adding to its steam electric generating capacity the Water Board in a limited way entered another public utility area—the sale of steam for heating and industrial purposes. When, in 1940, the Board was planning the steam plant addition, it was decided to acquire a large-size boiler to produce and sell excess steam since the Central Heating Company, the Eugene Fruit Growers Association and the University of Oregon had expressed an interest in purchasing steam for heating and other purposes. Subsequent investigation indicated that the potential market was small. But there were enough customers to warrant the Board entering into contracts to sell surplus steam after the larger boiler had been installed.

Throughout the Great Depression of the 1930s the Board was able to maintain and increase its sales of electricity and water. By 1934 there were 5,319 average customers on the water system and just over 7,000 customers on the electric system. Probably this seeming success in hard times was attributable more to the policies adopted by the Board than to the natural growth in the demand for these useful commodities. Electric and water rates were reduced a number of times during the period. Many extensions of water and electric service to consumers outside the city limits of Eugene were made. Advertising and in-person promotion were increased. Building of the electric load was encouraged by a cooperative appliance sales campaign. Little was overlooked and the results were gratifying.

Joseph W. McArthur; general superintendent-secretary August 7, 1932 - February 16, 1946. (EWEB files)

"ALL ELECTRIC HOME" RATE ADOPTED

Reductions in both general and special rates took place during the 1930s. The rates, rules and regulations governing water and electric service were revised in April, 1934 and in August of that year a new residential rate—the "all electric home" rate—was adopted. For homes with all major appliances using electrical energy, 7.5 mills per kilowatt-hour was charged for all use over 175 kilowatt-hours per month. In April, 1935 some further reductions were made in residential, commercial and power service rates. The next year, commencing May 1, 1936, rates were again reduced for electricity and rates for water during summer use were also reduced. The Board adopted, effective June 15, 1937, a special rate of 6.5 mills per kilowatt-hour for off-peak controlled storage water heating (at the same time discontinuing the $10 wiring allowance it had previously given for installation of electric water heaters). The last simultaneous reduction in both water and electric rates during this period was made effective February 1, 1939 with a final lowering of electric rates during the period on March 1, 1940. As post-war problems began to be considered, the possibility of increasing electric rates was discussed in order to enable the utility to catch up on the backlog of repairs and construction resulting from enforced neglect during the war years. On December 10, 1945 the Board decided against increasing rates, concluding that the growth of the system would increase sufficiently to generate required revenues.

The Board moved cautiously in making large amounts of electric power available to outlying districts. In December, 1936, when the Triangle Lake Rural Electrification Project Committee approached the Board with a request for a rate for wholesale power, the Board considered the policy implications. It decided that because of the rapid growth in demand for electrical energy in Eugene, it was best not to consider even a tentative rate but rather, encourage such groups to seek their supply of electricity elsewhere. However, after additional power was available from Bonneville and a new steam unit went into operation, the Board was willing and able to supply electricity in wholesale lots at a special rate. After two years of negotiation, the Board, in 1941, entered into a contract with the Lane County Electric Cooperative Association providing power at a wholesale rate for that part of its operations east of the Leaburg Dam (the Board having previously sold its distribution system in that area to the cooperative). In 1944 another contract was entered into with this organization for the sale of power for use in an area southwest of Eugene. Also in 1944, the policy with respect to rural extensions to individual customers was liberalized. Concluding that experience justified greater investment in such extensions, rules were adopted providing more free extensions and lessening the impact upon customers when charges had to be imposed.

The Board also entered the wholesale water business during the 1930s. In 1934 negotiations were concluded with a group outside the city with an agreement to construct a pipeline to the city limits where it would connect with a main of the College Crest Water District and to sell water to the district at a wholesale rate. That such extensions would be made only to the city limits was made clear in 1935 when the Board, acting on a request to maintain a small water main outside the city, reaffirmed its long-established policy of not constructing or owning water mains outside the city limits. Additional arrangements were made to sell to other water disticts willing to meet the Board's water mains at the city's boundaries in succeeding years. The Board agreed to sell to the South Willamette Street Water District and to a group living south and east of the city in 1937, provided that the districts would make satisfactory arrangements as to income, rates and conditions of service.

In 1939, 1941 and 1943 the Board entered into similar contracts with the Bethel Water District, the River Road Water District and the Glenwood Water District. Since it was serving water districts on almost all sides of Eugene, the Board began to consider serving the entire Eugene-Springfield area with water but took no action during the McArthur administration.

PROMOTION OF SALES VIGOROUS

Promotion of the sale of electricity and water by advertising was increased during these years. Early in 1933 the question of advertising to promote sales was discussed and the sum of $1,200 was allocated for the remainder of the year. Throughout the balance of the McArthur administration advertising continued to play an important role. In-person solicitation similar to that employed in the years of competition with the Oregon Power Company was not pursued. But in 1937 the Board authorized the employment of a home economist to encourage the use of electricity by demonstrating the many uses to which it could be put. Her employment was relatively brief as she resigned in August, 1939. The Board took no action to replace the home economist, indicating that it thought this kind of activity was not particularly effective.

The excess capacity available during the early 1930s again involved the Board in the sale of electrical appliances. In October, 1933 a study was initiated to determine the possibility of building load by providing a method of rental of ranges and water heaters. This resulted by early 1934 in a decision to set aside a sum of money to assist dealers in the sale of ranges. The Board purchased the electric ranges, charged dealers' cost plus 20 percent and collected payments monthly from purchasers through the regular electric bills. Initial results indicated success and the Board noted in April, 1934 that "so many ranges have been sold to date that it appears that the revolving fund is not adequate to finance and carry sales in the early stages . . ." An addition was made to the fund at this time and again in June of that year it was decided to finance the acquisition of heavy-duty ranges by fraternities and sororities at the University of Oregon.

A later evaluation of the policy of financing the sale of ranges led to the conclusion that it was not satisfactory in all aspects. In December, 1934 the Board directed that all ranges in stock be sold and that no more be ordered. It was also decided that there would be no more consigning of ranges to dealers and that jobbing and warehousing would be confined to water heaters. Load-building efforts were to be confined to a limited amount of advertising and an approved amount of financing for water heaters and ranges to be open to all dealers on all standard makes. While the Board was willing to continue free wiring of ranges and water heaters, it thought at this time that its cooperation with merchants did not require the handling of goods for them. However, in 1935 the utility was still stocking water heaters and selling them through dealers.

December 13, 1938; excavation begins on College Hill Reservoir No. 2 (EWEB files)

The free distribution of electric ranges to schools in Eugene began in 1933. On August 14 the Board authorized the placement of four ranges in the Eugene public schools for "advertising purposes." In 1935, noting the preponderance of gas equipment in these institutions, the Board authorized the purchase and distribution of electric plates for junior high and high schools, also giving ranges to junior high schools. The ranges came out of the stock which the Board had acquired as part of the cooperative program it had arranged with local appliance dealers. Eventually all of the public schools of Eugene and the parochial schools were given ranges by the Board. What had begun as essentially an advertising activity was beginning to assume the appearance of a community service program.

In addition to the involvement with the financing of ranges and water heaters, a further step was taken in furnishing the "all electric home" with the addition of air heaters to the list. A representative of the manufacturer of an electric air heater approached the Board in August, 1935, requesting that it finance in the same manner as it did ranges and water heaters the sale of small air heaters. After initially refusing—on the grounds that the price was not high enough to cover its overhead costs—the Board decided in early September, 1935 to give the plan a trial until January 1, 1936.

FINANCIAL HEALTH ENJOYED

Financial problems were few during McArthur's administration. Outstanding debt was retired without the necessity of issuing new bonds. Although some problems were experienced with collection of utility bills, there were sufficient earnings from the sale of electricity and water to meet the financial needs of the time. With loan funds available from the federal government on easy terms the Board attempted to secure some of this money, but without success. The principal problem, other than collections, was the matter of investment of surplus funds.

The interim power and light bonds due in 1936 and 1937 were called well in advance of the due dates. By May, 1934 sufficient funds were on hand to call all such outstanding securities with the exception of those already held in the sinking fund. During the remainder of the year purchases were made and in December, 1934 all of the interim bonds were retired.

Problems in bill collections during the Depression caused some concern. When customers did not have cash they paid in tax warrants and in "kind." In October, 1933 the Board had on hand slightly over $20,000 in warrants and anticipated the receipt of at least $40,000 more during the coming year. A policy was adopted of limiting the acceptance of warrants as much as possible and of banking or selling them when possible. At the same meeting the Board accepted as payment of an account a safe from a store which was going out of business.

Earnings and collections were sufficiently adequate, however, to pay for improvements during this time. The Board was tempted by the prospect that the Reconstruction Finance Corporation might lend funds on very generous terms. In April, 1933 General Superintendent-Secretary McArthur raised the question as to the advisability of making an application for Reconstruction Finance Corporation money for use in financing improvements to the water system. After some discussion it was decided that it would be better, whenever possible, to finance these improvements out of earnings rather than to borrow. Six months later the Board reconsidered and authorized the application for a loan of approximately $80,000 for a cover on the College Hill Reservoir and for new water mains. This was not granted and the Board proceeded, financing the improvements out of earnings.

INVESTMENT POLICY REVISED

The presence of a professor of law on the Board led to some change in the investment policy. Surplus funds of the organization had been invested in the bonds of a number of municipal corporations in Oregon, some of which were too small to qualify as satisfactory issuing agencies for public investment purposes under Oregon law. Upon the discovery of this situation by the Board in May, 1935 it divested itself of these investments and placed the funds in securities which it could legally hold.

Its relations with the community becoming ever more complex, the Board strove to bring some appropriate balance to its relationships with the various individuals and groups making claims on it. Customers who ultimately provided the funds for all of its operations were well-served. The City Council, under severe financial pressure, made greater demands on the utility operation. Relationships with the state and federal government became increasingly important, especially with the establishment of the Bonneville Power Administration. Relations with the Mountain States Power Company became increasingly competitive. Local groups, including business and labor, required continuing attention. An expanding work force necessitated the adoption of greater formality in employee relations. Finally, World War II brought special problems to be solved.

After the electric rate reductions which took place during the previous decade, the Board's attention was called to the fact that as of January 1, 1942 its residential-electric rates were lower than those of all other cities in Eugene's population class. In addition to enjoying such low-electric rates, Water Board customers benefited from other considerations extended by the Board. Beginning in 1935 the Board began participation in a "Eugene Beautiful" campaign by providing water at reduced rates during the summer months for residences, churches, fraternities and sororities. This policy was continued until 1945 when a water shortage led to its termination. Customer complaints were given prompt attention. When objections arose out of oversolicitation by business of new utility service customers, the practice of distributing the daily connect and disconnect list was discontinued.

The sale of water to individuals and industries outside the city limits posed certain problems. The Board preferred to deal with the water districts, but on occasion sold water to other parties when circumstances indicated that such action was desirable. A number of water main connections were made with industrial plants located outside the city limits—particularly lumber mills and plywood plants. Persons living along the canals and pipeline were permitted to tap them for irrigation and other purposes. In the sale of water to water districts the Board retained authority over rates to be charged the ultimate consumer to insure that they would not be higher than if the Eugene Water Board served the customers directly.

January 5, 1939; Construction proceeds on College Hill Reservoir No. 2 (EWEB files)

RELATIONS WITH COUNCIL TENUOUS

Relations with the Common Council of Eugene suffered during this period—a casualty of the Depression. The 25 percent discount on water and electric service to the city was continued until 1944 in the hope that this would assist in the retirement of the warrant indebtedness of the municipality (the Board being a large holder of warrants). In 1935 the Board granted the Council a reduction on the charge for electricity used in heating the city hall, which the Council, in seeming ingratitude, followed with a proposed increase of more than 100 percent in the rent paid by the Board for space in that building. In 1936 the Council made a large claim for monies it contended were owed by the Board. In late summer of that year the Council presented the Board with a claim for $81,000—$58,000 arising out of the Walterville power plant construction cost before the Board took it over in 1911 and $23,000 resulting from the appraisal by the Public Service Commission in 1917 of the Skinners Butte property, which the Council had taken off the Board's hands in 1914. The Council added accumulated interest to these amounts. The Board thereupon presented the Council with a counterclaim of $133,000 plus interest—for the value of the discount given the city for some seven years plus the difference between the charges actually made for fire protection, hydrant rental and reservoir maintenance, and those which the Public Service Commission had recommended. The Council then resolved that "both claims are stale and that there have been mutually reciprocal and helpful relations between the city proper and its Water Board during the years, and it is neither desirable or just that they should be ruptured by 25-year-old claims." The claims and counterclaims were then withdrawn and the discount to the city continued.

This apparent harmony between the two bodies was brief. The mayor and the Council, in September of 1938, proposed charter amendments whereby certain obligations of the city would be transferred to the Board. One proposal was for the providing of free water and lighting for municipal use, which the Board had historically opposed and which it continued to oppose as it stated in the voter's pamphlet on the issue. This amendment was defeated on November 8, 1938, although the Board was obliged by another proposal to make collections of the city's sewer charges and remit them to the city treasurer.

Due to lack of success in getting voter assistance toward the end of acquiring Board revenues through various means, the Common Council prosecuted its assault on utility funds in the state Legislature. Notwithstanding the opposition of the Eugene Water Board and other municipal utilities in Oregon, the Legislature in 1943 passed a law requiring payment by such organizations of three percent of gross annual electric operating revenues to the municipalities in lieu of taxes. When the Board learned of this legislation it voted to discontinue immediately the discount given to the city, but rescinded this action after the Council pointed out that this would jeopardize the city's financial condition since it had not taken such action into account in its current budget. The Board then decided to continue the discount through the fiscal year 1943-44, but that effective July 1, 1944, all special discounts would be discontinued.

Where money was not involved, however, the Board aided the city when possible. In July, 1938 the utility turned down a request for contributions for improving the playgrounds of the city, stating "the Board, while expressing its appreciation for the merits of this work, did not feel that it would be justified in appropriating revenues from the customers of the electric and water departments at this time no matter how worthy the purpose." During the following summer, however, the Board authorized the Playground Commission to use the cover of the new College Hill Reservoir as a playground.

The Board was not passive in its resistance to the attack upon its revenues in the legislative halls and joined other municipal utilities to forestall the passage of legislation permitting their taxation. In 1933 a contribution of $150 was made to the city of Ashland to assist in the preparation of a booklet opposing the taxation of municipally owned utilities. In June, 1939 a payment of $250 was made to the Water and

Light Department of McMinnville for part of the services of a special attorney during the preceding legislative session which "were very effective in preventing undesirable legislation affecting our municipal utilities." In 1941 the president and secretary of the Board were named as a committee to study possible future legislation on the subject of municipal-utility taxation. In February, 1943 when the attention of the Board was called to the fact that the bill proposing taxation of municipal utilities was sponsored by the Lane County delegation in the Legislature, it contacted this group and conferred with the legislators. The Board lost its fight, but to the extent that it had been able to delay the passage of this legislation, it had saved its customers many dollars. However, the cost of deteriorating relations with the City Council was very difficult to measure.

GOOD RELATIONS WITH STATE GOVERNMENT MAINTAINED

Although in disagreement with the Legislature on the matter of taxation, the Board had many constructive relationships with the state government. The Fish and Game Commissions were concerned with the effects of the Walterville and Leaburg plants on commercial and sport fishing and the Board aided in the construction of fish racks and rearing ponds. In 1941 it leased property at a nominal figure to the State Game Commission at Leaburg Dam for the building of trout ponds, although the state body was unable to guarantee that only Lane County waters would be stocked out of those ponds, as the Board—influenced by local sportsmen—had asked. In 1938, when a car left the highway and plunged into the pond at Leaburg Dam, the Board urged the State Highway Commission to put up guard rails and later shared in the cost of their erection. The Board had a special interest in this since at the time of the accident the attendant at the dam had opened the gates, when implored to do so by those witnessing the accident, in the hope that the driver might be saved. This action earned praise of those at the accident site but was condemned by fishermen and others downstream who were endangered by the sudden rush of water. Through cooperation with the highway agency the Board hoped to avoid a recurrence of this kind of problem.

President Roosevelt's New Deal created changing relationships with the federal government. Whereas it had previously been involved with relationships primarily with the activities of the Federal Power Commission, it now had a host of new relationships which were for the most part satisfactory. Although the Public Works Administration had denied a 100 percent grant in 1933 for the construction of the so-called Rennie Power Project, it later made a number of grants to the Eugene Water Board. Beginning in 1935, PWA money was secured to aid in the construction of water mains, construction of a new transmission line from Walterville to Eugene and in building the large reservoir on College Hill. The Works Progress Administration aided the Board in the period 1935-41 in improving the park at Leaburg Dam and in laying water mains—for which it furnished some materials as well as labor. The Board was also involved with the short-lived National Recovery Administration when it adopted in 1934 the NRA Code for publicly owned utilities drafted by the National Municipal Utilities Association.

February 14, 1941; Construction of Steam Plant Unit No. 2 (EWEB files)

BOARD PREPARES FOR WAR

As the New Deal waned and emphasis was placed on defense activities, the Board's relation to the federal government changed further and it now aided the national effort rather than being the recipient of federal aid. Even before Pearl Harbor, equipment was purchased to protect the utility's properties from sabotage. During mid-1941 chain-link fences were erected, security was strengthened at the Leaburg and Walterville plants and flood lights were ordered. In January, 1942 further measures were taken. Bulletproof protection for main-transformer banks was ordered and entrance to Water Board property restricted to employees and others on official business. The Board bought defense bonds and made deductions from the paychecks of employees who wished to buy bonds. A blackout switch was installed for the street lights of the city and provisions were made regarding canals, dams and powerhouses. Camouflage painting for the reservoirs was discussed and the water mains extended to the airport, where a school was established for training civilians for the Army Air Corps.

World War II also brought a temporary change in the Board's policy toward contributions. In 1934 $200 was given to develop a park area near the new Skinners Butte Reservoir and in 1937 the Board gave free electrical service for the Chamber of Commerce "Iron Lung Fund" dance at the armory. This was "contrary to general policy" and no other exceptions were made prior to 1941. During that year free electricity and water were furnished to Army troops staying overnight in the city at the fairgrounds. In 1942 free electric and water service was given to the United Service Organization (USO). The Ration Board ran out of funds in the summer of 1943 and the Board then gave it free light service until the end of the year. Contributions of cash were also made and the Red Cross was given $150 during 1943. Additional sums were given until the war was over, when the Board reverted to its established policy of no contributions of funds or services.

During the 1940s it became apparent that the federal government was going to play a stronger role in the development of power facilities on the McKenzie River and its tributaries. In April, 1940 the Board was advised by the Federal Power Commission that it should take steps either to perfect or withdraw its application for hydroelectric power development projects on the upper-McKenzie. The Board decided not to give up any of its power rights and instructed the general superintendent-secretary to investigate with state and federal authorities the matter of its rights and how best to protect and maintain its investment in power development and water rights on the McKenzie River. After this action had been taken the Federal Power Commission wrote in July, 1940 that it would withhold action for a reasonable length of time. Nothing further developed until June, 1945 when the Board, informed that the Corps of Engineers was considering building a dam at Quartz Creek, adopted a motion to apply for the first rights to develop and use the electric power that might be available at that site.

BPA ENTERS SCENE

The Bonneville Power Administration was a new federal agency destined to play a large role in the operations of the Eugene Water Board's electrical system. The Board commenced negotiations with BPA in January, 1938 for purchase of power, although a final agreement was not secured until the latter part of 1940 (due in part to the fact that the Federal Power Commission had to pass on the matter of rates). Some question was raised, chiefly by the editor of the *Eugene Register-Guard*, William Tugman, on the effect of the contract on "home rule." When the Board's attorney stated that the proposed contract did not in any way impair the jurisdiction and authority of the Board over its own rates and other matters properly under its jurisdiction, the Board authorized the signing of the contract on August 19, 1940 although it was not ready to receive service until December of that year. This contract made it possible for the Board to sell to Rural Electrification Administration cooperatives outside its previous distribution area.

Competition as well as cooperation characterized relationships with neighboring electric utilities during the period. Petitions had been received from certain customers of the Mountain States Power Company during December, 1933 asking that the Eugene Water Board serve them electricity. The Board thereupon authorized McArthur to negotiate with Mountain States Power Company for the purchase of such of its distribution system as might be necessary to serve the petitioners. The privately owned utility was unwilling to make such a sale and after repeated offers, the decision was made in February, 1936 to build distribution lines in some of the competitor's territory north of the Willamette River. Not all relations with Mountain States were competitive. In the early 1940s an agreement was reached with the investor-owned utility to sell or exchange surplus power, an arrangement resulting in profit to the municipal utility.

The Board very carefully considered its relations with local businessmen during this period. Dealers in electrical appliances were of considerable concern to the Board, as their activities influenced customers as well as the Board itself. To protect the interests of customers, some policies were changed and others adopted. In 1935 when attention was called to the possible injustice to the customer in paying $10 directly to the dealer as an allowance for the wiring and plumbing installation costs of water heaters, the Board adopted the modified policy of paying the amount directly to the customer or crediting the customer's account. The Board also believed that dealers were charging too high a price for water heaters. While buying water heaters from the Board for $51, dealers were in turn selling them to customers for $79.50 Thus, while the Board pursued increased use of electricity, it was mindful of the interests of its customers.

In 1936 the dealers approached the Board with a proposal which would have involved the utility in a much closer cooperative relationship with them. The Lane County Electrical Service Club asked the Board to employ a home economist, to advertise cooperatively with the club, to purchase ranges including warehousing and distributing them to dealers as needed, extend financing to appliances other than ranges and water heaters, appoint someone to act on the merchandising committee of the club and eliminate delays of remittances to dealers (the customers paid for the appliances on their utility bill with the Board transmitting the funds to the dealers). The Board declined most of these requests, including entering into cooperative advertising, the purchase of more ranges and the hiring of a home economist. However, payment arrangements were modified somewhat in favor of those dealers with good records, although it was further stipulated that purchase contracts in arrears were to be reassigned to the dealers.

In its relations with local businessmen other than the electrical dealers, the Board was chiefly interested in seeing that each got his "fair share." Although dealings with insurance companies as in the case of insurance on the steam boilers and turbines were direct, the policies were placed locally and a committee was set up in 1934 to determine the proper allocations of Water Board insurance business among local agents. When the issue of insurance placement was again raised in 1937, the Board decided to keep the insurance business distributed to some extent among the various agencies in Eugene rather "than have it all gravitate into the hands of one company due to low price." Also in 1934 the Board adopted a policy of distributing its funds among the various local banks in proportion to the capital and surplus of those institutions.

May 6, 1941; installation of turbine, Steam Plant Unit No. 2 (EWEB files)

CONCERN FOR LOCAL LABOR DEMONSTRATED

The Board continued to demonstrate concern for local labor during the McArthur years. The request for federal funds in 1933 for building a power project was made in part because of the large amount of unemployment in the Eugene area. Though the request was not granted, the Board decided in April, 1933 that some work be done on the grounds in the Leaburg area in order to create work for laborers needing it. When construction was being performed under contract in 1941, a criticism by local labor unions of the contractor for not paying prevailing union wages led to the Board's investigation of the matter and an eventual adjustment by the contractor.

The more important labor relations of the Board were those with its own employees and a number of changes were made between 1932 and 1946. After the wage reductions at the beginning of the period, there were few changes in pay prior to 1941. However, Board members Orlando Hollis and Percy Brown developed a new position and wage classification plan for employees which was adopted in July, 1933. This schedule arranged the titles of positions to indicate as nearly as possible the work performed with attendant wage levels.

Defense preparations in the war period itself brought wage problems to the Board. Numerous wage adjustments were made early in 1941 and when September and October brought additional requests for increases from employee groups, the Board authorized a study of salaries and wages of all employees in order that the matter might be handled uniformly. A new wage and salary schedule was adopted effective January 1, 1942 with some changes also being made in practices concerning leaves and holidays. Throughout World War II there were difficulties with wages due to the rising cost of living and the higher wages paid in logging and other war-related industries. The Board did not adopt any policy of adjusting wages on the basis of cost-of-living increases nor did it make any general increase in wages, relying instead on individual wage adjustments. In the closing days of the war studies were made of comparable wage and salary schedules and meetings were held with a committee representing the employees. However, these meetings were generally unproductive.

CLEAR PERSONNEL POLICIES LACKING

Changes in hours worked during the years 1932-46 reflected the Depression conditions and the increased influence by employee factions. In March, 1933 the number of shifts at the steam plant and the water filtration plant was reduced, with no one to be laid off but with work assignments reorganized. In 1940 the employees began to agitate for a reduced work week. Representatives of the safety committee presented to the Board in April of that year a request for a 44-hour week as a substitute for the existing 48-hour schedule. Nothing further developed until 1944 when a committee representing the Board's employees again approached that body asking a 44-hour work week. On July 1, 1944 the Board granted the request for a 44-hour work week with the understanding that employees would work 48-hours at straight time, if necessary, with overtime pay applicable only to work in excess of eight hours a day or 48-hours a week.

When the new wage schedule was adopted in December 1941 a number of changes affecting leave were made. The Board voted to discontinue the practice of paying the employees the difference between the amounts paid by the State Industrial Accident Commission and the salary of the employee for lost-time accidents. Also beginning January 1, 1942 the Board limited sick leave to one week and made it available only to monthly employees. At the same time all monthly employees were allowed the equivalent of seven holidays and the practice of allowing two weeks' annual vacation with pay was continued. Upon presentation of a request by an employees committee in 1943, the Board modified these policies to make sick leave cumulative and to give daily employees a vacation allowance. When conscription began to affect the Board's employees a policy similar to the one of World War I was adopted; that is, any employee entering military service would have his job held open for him with any replacement in his position to be temporary.

October 27, 1941; Joe McArthur (left) and Hugh Currin inspect new steam plant unit. (EWEB files)

During this period some changes were also made in insurance coverage for employees. There was some increased coverage with office employees also coming under the State Industrial Accident Commission in 1933. The Board was not willing to contribute to any additional programs, however, although it permitted, in 1933, an insurance firm to solicit employees for voluntary participation in health and accident insurance. In 1935 two insurance firms asked that the Board engage in group plans of health and accident insurance, but the members of the Board were unwilling to accept either plan even though its part in the proposal would be limited to administering payroll deductions. Throughout these years participation in group insurance was limited to that covering the lives of the employees. Initially the Board continued to pay the employer's share of group life insurance premiums of employees in military service, but adopted a policy in October, 1941 requiring that all employees entering such service be required to relinquish group insurance on the grounds that government insurance was available.

Retirement plans were considered during this time, but no action was taken. As early as 1935 an investigation of some form of retirement plan for the Board's employees was undertaken without any immediate result. Ten years later the Board's attention was called to the fact that the Legislature had just passed a bill providing a state-wide plan for retirement of public employees in Oregon, and that the Board would have the option of rejecting the plan provided it did so before May 1, 1946. Unable to make a decision, the Board took no immediate action but eventually adopted the plan.

Throughout the period of 1932-45 the Board and administration acted with paternalism and apparent lethargy with respect to employee relations. An attempt was made to take into account what was thought to be the best interests of the employees. However, issues were often treated on the basis of individual merit rather than established policy guidelines. Even when issues arose which could have led to the establishment of constructive policies, the Board did not act. As a consequence, the organization lagged behind other employers in putting into effect good personnel practices. Increasing employee activity during the latter part of the period portended eventual formation of an employee union. This probably would have become reality had later Boards and administrations failed to give proper attention to employee matters.

While unresponsive as it might have been to the personnel management needs of the employees, the Board did look to the future with respect to its chief administrator. In January, 1945 it declared that the system had grown to such an extent that the general superintendent-secretary should have a duly recognized assistant, and thereupon appointed Ray B. Boals to that post. One year later J. W. McArthur asked to be relieved as soon as practicable, giving as reasons his advancing age, failing health and the pressing need of a younger man to carry on the work of general superintendent-secretary in the new era of expansion lying ahead for Eugene. The Board adopted the recommendation providing that McArthur be relieved of his duties as of February 16, 1946 and resolved that Ray B. Boals be appointed as of that date.

CHAPTER FIVE
A New Surge Forward

THE EUGENE *REGISTER-GUARD*, in the early months of 1946, carried stories reflecting the hopes and expectations of the post-war era. There were still somber reports of the death of young men of the area who had been missing in action since early in the war or were known to have been taken prisoner by the Japanese. The University of Oregon was experiencing its first heavy surge of veteran enrollment.

Ray B. Boals had been employed by the Eugene Water Board as mechanical engineer since 1938. His appointment as assistant general superintendent-secretary in 1945 demonstrated the Board's confidence in him as a successor to Joe McArthur. A graduate of Oregon State College with a degree of bachelor of science in mechanical engineering, Boals took graduate work at Carnegie Institute of Technology and was employed for a period of time at Duquesne Light Company of Pittsburgh. He also taught engineering at Oregon State for nine years and served as light and water commissioner and city engineer for the city of Fairbury, Nebraska. His education and experience seemed ideal for the top administrative position at the Water Board.

When Boals took over on February 16, 1946, the Board declared that he was to have full power and control over the management of the electric and water utilities, responsible only to the Board itself. Boals served in this position until he resigned in March, 1957 after exceeding the age of 65.

At the beginning of his 11-year tenure as chief administrator, Boals presented a 10-year municipal light and water expansion program for Eugene and the vicinity which would double existing capacity. He presented his program at the May, 1946 meeting of the Board. Major projects in the 10-year plan for the electric utility included rebuilding and extending the distribution system, an addition to the steam-electric generating plant, installation of a more efficient unit at Walterville and the addition of a second generating unit in the Leaburg plant. Water system improvements proposed by Boals included two new reservoirs, a new filtration and treatment plant and a new water-supply line from Hayden Bridge.

Ray B. Boals, general superintendent-secretary February 16, 1946 - June 30, 1957. (EWEB files)

NEW GENERATION ADDED

In 1947 some consideration was given to building hydroelectric generation near Eugene—at Hayden Bridge and near Goodpasture Bridge. These concepts were rejected, however, and the Board proceeded with plans to rebuild the Walterville plant, add a third unit to the steam plant and a second unit to the hydro plant at Leaburg. This construction program was completed in 1950.

During the period from 1940 to 1950 water customers increased from 6,090 to 9,013, an increase of 48 percent. Electric customers increased from 9,520 to 18,577, an increase of 95 percent. The reason for the greater increase in the electric system customers was that the Water Board provided direct service to electric customers living outside the city limits while water users outside the city were served through water districts and, for purposes of accounting, each district constituted a single customer.

UPPER-McKENZIE FAVORED

As the electric load grew rapidly in the post-war years, it was soon apparent that additional capacity would have to be added beyond that recommended in 1946. Although the Board had acquired power rights at the Rennie Dam site in the 1920s and had later acquired the property itself, it gave up these rights in the years immediately following World War II. In 1950, however, the Board again began to examine the possibility of building an additional hydroelectric plant at other sites on the McKenzie River. The alternatives to this were to increase power purchases from Bonneville Power Administration or to build yet another steam plant. Preliminary studies indicated that a new hydroelectric plant on the upper-McKenzie River would best suit the needs of the system and in 1951, a preliminary permit for an upper-McKenzie project was secured from the Federal Power Commission.

Studies of possible locations narrowed the field of choice to two locations—one at Carmen Creek and the other at Beaver Marsh. The results of further investigation led the Board to decide in 1954 that Beaver Marsh was the best location for new generating facilities. In January, 1955, after the decision was made, the Board authorized an expenditure of $106,000 for the preliminary work, such work to be discontinued if either the Federal Power Commission or the State Engineer refused to issue a license.

VOTERS REJECT UPPER-McKENZIE PROJECT

After the state approved the project and the Federal Power Commission issued a license in January, 1956, all that remained to be done was to convince Eugene citizens that they should approve a measure authorizing the project and providing funds for its construction. Plans were made for Board employees to talk to service clubs and other groups, explaining the project. Favorable news stories were forthcoming and an additional expenditure of funds was authorized for advertising the Board's views on the proposal. These plans and programs seemed to have been for nothing when the voters turned down the measure on March 27, 1956. The vote was 1,846, yes; 2,238, no.

There was a great amount of controversy surrounding the Beaver Marsh project. The Save the McKenzie River Association, National Parks Association, Federation of Western Outdoor Clubs, the Isaac Walton League and the Seattle Audubon Society were among organizations opposing the measure. Oregon Senators Richard L. Neuberger and Wayne Morse introduced a bill in Congress which would have revoked the Federal Power Commission's license that it had granted. Going to Washington to testify in favor of the bill were Karl Onthank of Eugene and former Oregon Governor Charles Sprague. Testifying against the bill were Board Counsel Windsor Calkins, Deputy Director-Operations Byron K. Taylor of the Water Board and Fred Brenne, Eugene Chamber of Commerce manager.

In the election on March 27 the vote against the measure may have been most heavily influenced by the *Register-Guard* editorial which appeared on March 26. While praising the Water Board generally and in particular with respect to the excellent job of planning which had been accomplished, the *Register-Guard* simply concluded that the scenic beauty of the upper-McKenzie was of primary importance and the project should not be built.

PARTNERSHIP SOUGHT

At the same time that it was pursuing the campaign to build generating facilities on the upper-McKenzie, the Board was attempting to form a power-generating partnership with the U.S. Army Corps of Engineers. In June, 1953 the general superintendent-secretary was authorized to inquire if and under what conditions it would be possible to purchase, own and operate the generating facilities to be installed at Dexter Dam on the Willamette River. Nothing resulted from this inquiry, but in July a step was taken which almost put the municipal utility into partnership with the federal government in the development of power resources on the McKenzie River.

On July 13, 1953 the Board authorized filing for an application of water for power purposes on the south fork of the McKenzie. The Corps of Engineers had abandoned its plans for a large dam on the main stream at Quartz Creek and was planning a number of smaller installations on the tributaries. One of these was at Cougar Creek and the Board wished to obtain water rights at this site so that it might own and operate that part of the multi-purpose installation which would generate electricity.

The Cougar Dam project was not considered as a possible alternative to Beaver Marsh, but rather as a supplement to it since there was some indication that the Cougar Creek development could be economically tied to the Board's plans for eventual development of generation on the upper-McKenzie.

By December, 1953 additional steps had been taken toward involvement in the Cougar Dam construction. A brochure was prepared for Congress and others showing the Board's interest in power development at this location with a proposal made to participate in the construction and to underwrite the cost of installing generating facilities. Financial institutions were asked for their opinions concerning financing the Cougar Dam and Beaver Marsh projects. Board interest in these sites was strengthened when a consulting engineer reported that an investigation of sites on the Siuslaw River did not reveal any location which would compare from a cost standpoint with the other two projects.

The support of the Eisenhower administration was secured for the partnership concept in early 1955. On February 15, 1954 the Board made additional filings for the Cougar Dam project and authorized other necessary steps. The result of the activities of the Board and management of the utility was a recommendation by President Eisenhower in his January, 1955 budget message to Congress that an appropriation be made for the Cougar project, provided that it was accepted as a partnership concept.

Unfortunately for the plans of the Board, a political snag was hit in the quest for control of power generation at Cougar Creek. Senators Morse and Neuberger opposed partnership generally and the specific proposal. Finally, in May, 1957, the Board decided to withdraw its application from the Federal Power Commission for a license to generate power at Cougar Dam and the matter was permitted to expire.

The rapid increase in demand for both water and electricity in the post-war period taxed the ability of the utility to construct distribution facilities. In September, 1946 a report on the progress of electric and water connections disclosed that despite some improvement in the transformer situation, it was impossible to provide relief to a number of areas where larger transformers were badly needed to expand the use of electric ranges and water heaters. At the same time the water department superintendent indicated that there was a 30-day delay in connecting water service. Inability to secure an ample labor supply had led to the use of the service crew for laying mains rather than for installing services. This situation was relieved somewhat in December, 1946 when provision was made for installing mains under contract—the first time this means had been employed.

The need for additional water storage capacity also led to new construction. On April 10, 1950 the Board authorized the construction of the College Crest and Fairmount high-level reservoirs. In June, 1951 negotiations were begun for property at the top of Gillespie Butte for an additional reservoir.

With the extensive growth in the utility system a need arose for new office, warehouse and shop space to support the main tasks of the organization. The Eugene Woolen Mill Company owned desirable property on the Willamette River next to the Board's steam plant site. In June, 1948 the Board began negotiations to purchase three acres from the company. An agreement reached shortly thereafter provided for the possession of all the property by the beginning of 1950. On September 11, 1950 approval was given to begin planning for building both a new warehouse and a new office building on the former Woolen Mill property. However, in February, 1951 the concept of separate buildings was discarded when it was decided to build a combination office and warehouse, since this move would permit moving the staff out of the crowded quarters existing in the city hall at that time and into the new building at an earlier date. In July, 1951 the Board authorized a new shop building to complete this construction program.

During the immdiate post-war years the Board was again confronted with the problem of what to do about extending water service outside the city limits. The existing policy of selling only to water districts was brought into question by an inquiry in the summer of 1947 from the Mountain States Power Company as to the rate to be charged them for water from the proposed Hayden Bridge Filtration Plant. That company intended to retail the water in the city of Springfield, where it also provided electric service. The Board decided to defer decision until completion of the plant, although the thought was expressed that the rate should not be lower than that charged the water districts. In March, 1948, when the Chamber of Commerce asked for a statement of policy on extending water service outside the city—the Chamber being especially interested in the Seneca industrial area—the Board decided that the policy of selling only to water districts should continue. This policy was reaffirmed in January, 1950 when a contract with the Rainbow Water District was approved.

RATES GIVEN ATTENTION

Higher costs of construction of additional water facilities led inevitably to higher water rates after World War II. One of the first indications of this trend was the decision in June, 1946 to discontinue offering special summer water rates because of the costs of the heavy expansion program. Although rates were not immediately increased, by the summer of 1950, the water department was operating at a loss, necessitating a new and higher schedule of water rates effective the first of September of that year. Experience indicated that this was inadequate and on July 1, 1955 water rates were again increased.

With some exceptions, the general trend of electric rates continued downward after World War II. In November, 1955 the Board discussed the existing rate structure, concluding that the rates should be reduced by at least 10 percent. In December it adopted lower rates for residential service and all-electric residential and farm service. Commercial service rates were reduced effective February 1, 1956 and in December of the same year, yet another reduction in rates to residential and commercial customers was approved.

Electric rate increases which were imposed during this period reflect the Board's thinking on load configuration. In May, 1948 the rate for electric space heating was increased with the purpose of discouraging what was thought to be an undesirable load. In November, 1952 this policy was reaffirmed. In a revision of residential rates some reductions were made in lower consumption levels of electricity. However, a rate of 1.6 cents per kilowatt-hour for consumption over 1,200 kilwatt-hours was established, while only one cent was charged for consumption between 100 and 1,200 kilowatt-hours. This structure was designed to penalize heavy residential consumption associated with space heating. At the same time the flat rate for water heating was abolished. One year later, in 1953, the six mill penalty for electric space heating was eliminated based on new thinking that such consumption should be encouraged rather than discouraged because it was then considered "good load."

Based on a study initiated in December, 1950, the general superintendent-secretary, on February 12, 1951, told the Board that the existing power rate for industrial customers was significantly less than rates in Portland as well as those of other electric utilities in the surrounding area. No action was taken until 1954 when new rates were established effective July 1. These rates contained a combined commercial-industrial rate which had the effect of decreasing the commercial rate and raising the industrial rate. Such a change was thought to make a more equal distribution of charges whereby the industrial customers would not be subsidized at the expense of the residential and commercial customers.

Electric ratemaking historically had been a complex process. In 1892 a British rate expert, Dr. John Hopkinson, proposed the demand and energy form of rate. Another Englishman, Arthur Wright, in 1896 proposed the hours-use rate. Both of these rate structures take into account two components. One component is the energy or commodity actually used during the billing period. This takes the form of the kilowatt-hour representing the use of a kilowatt over a period of one hour. The other component is demand which is represented by kilowatts and, in the ultimate sense, indicates that part of the electric system which is called upon by a particular customer. Demand is billed on the basis of the highest level of kilowatt demand during the billing period. Essentially the theory behind this practice is that the electric system must be such that it can meet the highest or peak loads at any point in time. Therefore, the customer who is billed on a demand as well as an energy basis must pay his share of the cost of construction and maintenance of the total electric system. The same holds true for water system customers even though a clearly identified demand charge may not appear on the billing statement. In all utility systems a commodity charge must be imposed as well as a charge for the production, transmission and distribution of that commodity. It is for this reason that ratemaking can be extremely complex when performed in the interest of equity among customer classes.

The changes in electric and water rates during this period were made for different purposes. Ratemaking in the electric department was chiefly aimed at the improvement in the configuration of the electric load as well as establishing more equitable rates among the various classes of customers. In the water department, however, rates were increased because of financial stringency. A 1948 study had indicated that unless rates were increased the department would be at deficit by about $1.3 million at the end of 1949. The general reduction in electric rates and increases in water rates reflected a policy statement adopted by the Board in 1946:

> Although rates must be kept as low as possible, they must at the same time be high enough to provide funds for retirement of all debts when due, for replacement of equipment as it wears out, and for new capital to pay for extension and enlargements of the physical plant whenever demands for water and electricity increase.

"EWB" CHANGED TO "EWEB"

Because new capital was needed for purposes other than extension and enlargement, the Board proposed a number of new bond issues between 1946 and 1957. In the latter part of 1948 needed capital improvements in both departments of the utility prompted the Board to discuss additional bonds. On December 27, 1948 a resolution was approved for a charter amendment enabling the Board to issue $3 million in bonds. Coupled with this was a proposal permitting the utility to borrow money in an amount up to $500,000 on a short-term basis. Both of these financial amendments were approved by the voters early in 1949. On the same ballot was another measure which was to more accurately reflect the nature of the operations of the organization. The Eugene Water Board was henceforth to be known as the Eugene Water & Electric Board. On March 28, 1949 the Board decided to sell $2 million of the authorized securities and the final $1 million in bonds were issued in 1950.

When the Board decided to extend electric service into its private competitors' distribution territory in 1952, additional funds were required. In January the Board had discussed securing authority through a charter amendment to issue revenue bonds on its own motion, but it postponed a decision on this matter. To meet financial requirements as-

September 24, 1948; 45 inch-line on slope of Vitus Butte. (EWEB files)

August, 1949; construction of Hayden Bridge Filtration Plant. (EWEB files)

sociated with the acquisition of facilities in Mountain States territory, the Board sought and won voter approval at the November 4, 1952 general election for the sale of $1 million in bonds—to be used for improvement, acquisition and extension within and without the city.

After the voters decided against the Beaver Marsh project on March 27, 1956, which tied the authority to issue $8.5 million in bonds to that permitting the construction of the facilities, no more bonds were issued during the administration of Ray Boals. The defeat of the Beaver Marsh proposal appeared to be more a consequence of opposition to the construction of the project at the proposed location than to the amount of money involved, since it was provided that "in no event shall taxes be levied on the property in the city for the payment of these bonds," the first time this provision had been contained in a bond issue.

RELATIONS WITH OTHER ENTITIES BECOME COMPLEX

As the municipal utility expanded in size during the post-war period, its public relations problems and general activities became more complex. The issue of how best to serve water customers—including the question of fluoridation—proved a vexing one. Competition was renewed with a private power company serving the area and the Board in turn was faced with competition from new publicly owned utilities. At the same time it extended its cooperation to other electric utilities, both public and investor-owned. Local business and community groups continued to make demands on the Board. The friends of the water and electric utility were of continuing importance, especially those on the McKenzie River. Relationships with governmental bodies grew more complex with the city of Springfield and school districts outside Eugene being added to the groups with which the Board already had dealings.

The city of Springfield first raised the issue of what to do about the nature of its community water system after World War II. In February, 1946 some members of the Board met with the Springfield Water Committee and reported that that body wanted the Eugene Water Board to take Springfield's water requirements into account in making future plans. With the annexation by the city of Eugene of areas which had been served by water districts to which the Board had sold at wholesale, the question of who should serve the customers and at what rates was raised. The Board eventually made its own connections with these customers at somewhat lower rates than those charged by the water districts. In May, 1955 a meeting was held between water district representatives and the Board staff, during which was discussed, among other things, the possibility of creating a metropolitan water district. No Board action was taken toward this goal, however.

In 1951 the Board was asked to investigate fluoridation of the water supply. On May 14 of that year the Board agreed that the general superintendent-secretary should request an enabling ordinance by the

1950; pumping and valve system, Hayden Bridge Filtration Plant. (EWEB files)

1950; operator's control station, Hayden Bridge Filtration Plant. (EWEB files)

Council as the first step in fluoridation of water. After receiving the request the City Council referred the matter to its Health Committee for investigation and recommendation. More than a year later—in August,

1952—the response of city officials indicated that they preferred a delay until other cities had had more experience with fluoridation. This probably was a consequence of the unfavorable reaction of some individuals and groups to fluoridation. Both city officials and Board staff had received letters and telephone calls from parties opposed to the measure. Thinking that the matter ought not be laid to rest without an attempt to educate the public, the Board authorized on June 8, 1953 the use of newspaper and radio advertising for this purpose. This was followed in September by a survey of water customers. When slightly over one-half of the returns indicated opposition to fluoridation, the matter was tabled until public interest might prompt further consideration.

While the Board preferred to rely on its own resources for electrical energy, it took increasingly more from Bonneville Power Administration. A new contract was drawn with that agency in 1946 and the succeeding year brought an amendment to the agreement increasing the amount of firm power under contract. After the failure of the Beaver Marsh proposal in March, 1956, the Board decided not to submit the matter again in the November election of that year and in December, 1956 a new 20-year contract was drawn with BPA to insure adequate power supplies.

Other sources of power were acquired in addition to Bonneville. As producing units in the forest products industry grew larger, they burned waste in boilers using the steam for heating and the generation of their own electricity. The Board had earlier made an agreement to purchase surplus power from the Giustina Lumber Company. Negotiations were also underway at this time with the Weyerhaeuser Timber Company to purchase excess energy from that firm if and when it built a plant in the Eugene-Springfield area.

PRIEST RAPIDS POWER ACQUIRED

Much more important than the electrical energy purchased from lumber firms in the area was the arrangement made with the Public Utility District of Grant County, Washington. Grant County PUD was in control of the power to be generated at the Priest Rapids facilities on the Columbia River and sought purchasers for the energy. At a meeting in Portland during May, 1955 the Board staff was led to believe that it was not very likely that Grant County PUD power would be offered to the Eugene utility. However, in October of that year Grant County PUD offered the Eugene Water & Electric Board a proposed power sales contract. Although the Board could not immediately use all of the power offered, it secured the maximum available to it—with the surplus beyond its immediate needs to be sold to Portland General Electric Company and Pacific Power & Light Company. The 50-year contracts were negotiated in May, 1956 although it did not appear that the power would be available until 1960. It was also thought that this source might represent the lowest-cost power available in the future.

With additional power the Board was able to expand its marketing of electricity into areas not previously served by it. In July, 1949 Boals suggested opening negotiations with Mountain States Power Company asking for the purchase of that company's distribution system lying between the Willamette and McKenzie rivers north of Eugene and Springfield. This was justified, it was thought, on the grounds that the area would eventually be supplied with water from the Eugene system. Mountain States responded that the territory was not for sale provided that the Board did not plan to enter the area as a competitor. But the Board did have such plans and put them into effect in 1951, asking prospective customers in the territory in question and in the Glenwood district to sign petitions if they wanted to have service from the Eugene system. Early in 1952 the Board again asked Mountain States if it wished to sell, pointing out that if it was not possible to acquire distribution lines by purchase, the Board would go into the area in competition in order to protect its interests. When the private firm decided not to sell, the customer-owned utility began taking applications for service and preparations were made to serve customers in competition with Mountain States. The Board had an interest in serving the area east of Springfield and so advised persons living in that location. To further exert influence, the Board asked the City Council to file condemnation proceedings against Mountain States Power Company for the acquisition of certain parts of its distribution system ". . . necessary for the economic administration, use and functioning of the City of Eugene's electric system." The Council obliged the Board in this request.

Condemnation proceedings went forward and a judgement in favor of the Board was rendered by Lane County Circuit Court Judge Frank B. Reid on August 30, 1954. But it was threat of pending judgement, not the judgement itself, which hastened negotiations. The Board had learned early in 1954 that the Mountain States properties in question were in the process of being acquired by Pacific Power & Light Company. A meeting was arranged with PP&L representatives in July and in August, prior to the date of Judge Reid's action, an agreement was reached under which EWEB would acquire those facilities lying, in general, between the Willamette and McKenzie rivers as well as those in School District No. 4 and in the Glenwood area. The sum paid was slightly in excess of $1.22 million.

Although Lane County Electric Cooperative was not a direct competitor, the Board considered acquiring some of its properties during the early 1950s. In May, 1951 the question of possibly purchasing the cooperative's distribution system on the upper-McKenzie area was raised (this system had been sold earlier by the Board to the cooperative). No action was taken at this time and when the question again arose in early 1953, the cooperative was not disposed to sell, although it was thinking of the possibility of selling some lines south of Eugene.

The Board itself was confronted by competitors encroaching on what it regarded as its territory. In 1952 the Blachly-Lane Electric Cooperative began serving customers in areas already served by the Eugene utility. After some discussion of the matter in February, the Board decided that no advantage would be gained by pursuing the issue locally but that a complaint should be lodged with the Bonneville Power Administration—which supplied the cooperative with energy—and if no action were forthcoming, to take it to members of Congress.

NATURAL GAS ENTERS COMPETITION

Not only was there competition from other suppliers of electricity, but from gas companies as well. In January, 1955 the Board considered adding an employee to promote the use of electric energy for heat after it was announced that natural gas would shortly be supplied to the Willamette Valley. While the Board also discussed the use of cash subsidies to customers to pay part of the cost of rewiring homes so that electric heat might be used—as was being done in Tacoma and Seattle—the Board decided to take no action at that time.

The organization of a municipal-electric utility in Springfield brought a new competitor to the scene. In the summer of 1949 the city of Springfield asked the Eugene Water & Electric Board to "wheel" Bonneville power to the distribution system which Eugene's sister city proposed to establish. The Board agreed to do so and served Springfield by this means until 1952. After Springfield had its own connections with Bonneville it solicited customers in areas being served by the Eugene utility, despite an agreement to the contrary. The Board took counter measures in January, 1953 but the issue was still not resolved in 1955 when counsel for the Eugene utility submitted an opinion on the validity of the boundary agreement made between the city of Eugene and the city of Springfield dated July 21, 1949:

> It is our opinion that this agreement is still in effect and will continue to be in effect so long as the two cities own and operate electric distribution facilities if the same is not revoked by subsequent acts or a subsequent agreement between the parties.

The Board also suffered indirectly from the entering of the Springfield municipal utility into the electricity business in the Eugene-Springfield area. The Springfield organization was embarking on a campaign against Pacific Power & Light in Springfield. When PP&L chose to fight, it secured authority in 1955 to reduce the rates in its service area between the McKenzie and the Willamette rivers to the rates charged by the Springfield municipal utility. Since these new rates would be lower than the rates charged by Eugene outside the city (city rates plus 10 percent), a modification of the Board's policy was indicated although nothing was to be done about the surcharge until several years later.

Not all of the Eugene utility's relations with Mountain States Power and other utilities in the area were competitive. During December, 1946 and the first three months of 1947 the Leaburg power plant was isolated from the Water Board system and connected directly to Mountain States Power and energy was sold to the private firm when it had a special need. A good price was secured for the power, initially seven mills per kilowatt-hour, and after it was discovered that Seattle and Tacoma publicly owned utilities were selling energy to Puget Sound Power & Light Company at 10 mills, the price was raised correspondingly. The Board also realized income from wheeling Bonneville power to REA cooperatives.

1950; Hayden Bridge Filtration Plant in operation. (EWEB files)

The Board also cooperated with public and private organizations in the electric utility field through various associations. Ford Northrop of the Eugene Water & Electric Board staff was elected president of the Northwest Public Power Association in 1950. The Eugene utility had been a member for a number of years and now assumed a leading role. During the succeeding year Northrop represented the association and others in hearings conducted in Washington, D.C. on the delivery of Columbia River power to California. In March, 1955 the Board became a contributing member of the Pacific Northwest Utilities Conference Committee (which also included privately owned utilities) and agreed to remain a member so long as both public and private utilities were represented and the organization was of value to the Board.

Demands and activities of local business also required Board consideration. After discussions with Eugene retailers of electrical equipment and supplies, the Board adopted in 1950 the policy of not selling materials to electric customers except where they were not available from local retail dealers. In 1951 a Springfield electrical contractor asked the Board to stop two of its employees from doing electrical installation work as a sideline. This request was refused. The Board held that so long as outside activities of its employees did not affect job performance, it could not restrain them.

Local fuel dealers met with the Board in December, 1953 to protest the proposed reduction in electric rates, as such a move would encourage the use of electricity for heating and thereby damage their business. The Board listened but made no promises. During 1954 a Lane County unit of the National Electric League was formed consisting of electric appliance dealers, electric contractors and utility companies to promote adequate wiring of homes. The Board supported the organization by approving the payment of $600 per year for advertising and publicity. The Board also voted in 1955 to discontinue the practice of supplying lists of new customers for the use of business firms in soliciting business when local business firms began to abuse the privilege. This same action had been taken by the Board in 1942. Apparently the distribution of customers lists had been renewed at some point between 1942 and 1955.

BOARD SUPPORTS CHAMBER OF COMMERCE

The Board recognized the value of the Eugene Chamber of Commerce and expanded its support for that body in the post-war period. The chamber approached the Board in 1946 asking for a contribution of $50 per month. This request was declined although a contributing membership was maintained. In 1951 the chamber again requested an increased participation, this time asking $1,200 a year. The Board felt that it should not be the largest contributor and so authorized the annual sum of $750 to be given beginning January 1, 1952 (in August, 1953 the contribution was raised to $1,200 to begin in 1954). In 1952 a special request by the chamber for $3,500 for an industrial survey prompted the Board to authorize the purchase of 300 of the printed reports at $5 each in lieu of the donation.

Community groups other than businessmen also claimed the attention of the Board. The chairman of the minorities committee of the League of Women Voters of Eugene suggested in 1951 that the Board employ one or more Negroes in positions for which they were qualified. Exhibit space was purchased at the Lane County Fair during these years, both to support the fair and also as a public-relations measure. Staff members were asked to speak before such groups as the Rotary and Kiwanis, opportunities which were welcome when the Board wished to publicize and secure favorable public opinion with respect to Beaver Marsh and other proposals. Contributions to charitable organizations had been discontinued after the war, but the question was again raised in 1953 when the Board was solicited for a $1,000 subscription for the proposed McKenzie-Willamette Memorial Hospital in Springfield. The Board was willing to make this contribution, if legally possible, as a means of providing faster emergency service to its employees working and living on the McKenzie River. However, Board counsel gave the opinion that there was no authority to make such a donation, declaring "a donation of this kind would inaugurate a policy that would later vex the Board." The matter was then laid aside.

Although not in the community of Eugene, the McKenzie River Protective Association was destined to play an important role in the post-war activities of the Eugene Water & Electric Board. In May, 1952 the Board received notice that this group was calling a meeting at McKenzie School to conduct a "hearing" on EWEB plans along with other matters of discussion. The continuing activities of this group prompted the Board to adopt in February, 1954 an educational program to fully inform the public of its plans. Special attention was paid to educating the residents of the upper-McKenzie area regarding the Beaver Marsh and Cougar Dam projects. In July, 1954 a special rate was made for electricity sold to the Lane County Electric Cooperative, partly because the account might be lost to Bonneville but also to secure the cooperation of people in the upper-McKenzie area. A final step was the opening of Water Board Park at Leaburg in 1955 (the McKenzie group had as early as 1946 urged the Board to develop the area for recreation). In spite of these measures, the McKenzie River Protective Association remained firmly opposed to Beaver Marsh and was a factor in its defeat.

CONTRIBUTION VOLUNTARILY INCREASED

The desire of the City Council for a larger share of the funds of the Board clouded relationships between the two city bodies in the years following World War II. The Board attempted to improve relations with the city in 1948 when it agreed to build a structure at the airport to house a lighting unit and fire truck (with the city to repay the cost and to pay a monthly charge for equipment and service). In 1950 a more liberal plan was adopted with respect to street lighting although the Board was unwilling to change its policy with respect to hydrant rental and fire protection. Matters became worse in 1952 when the Council did not make prompt payment for fire hydrant costs and the Board stated that these bills must be paid before other expenditures in favor of the city could be justified. A joint meeting of the two bodies in January, 1953 produced no agreement, with the Council stating that it was being charged excessively for fire hydrant use and that the Board should assume ownership and maintenance of the fire hydrants. The Board disagreed, but made a study of the cost and adjusted charges somewhat in May, 1953. The Council was not satisfied, however, and proceeded along a different course, asking the state Legislature to increase the amounts which municipally owned utilities had to pay in lieu of taxes. Before the legislative issue was resolved, the Council asked the Board in May, 1955 to voluntarily increase its payment from three percent to six percent of its gross electric revenues. In its meeting the next month the Board consented to the request provided that ". . . surplus money . . . be used in reduction of property taxes within the city of Eugene." This action, approved for the calendar years 1955 and 1956, was extended during 1957 contingent upon action which might be taken by the Legislature, then in session.

The Springfield school district also sought money from the Board in lieu of taxes. The basis of this request was that some of the Board's employees lived in tax-free housing supplied by the Board with the children of these employees being provided educational services by School District No. 19. As a matter of compromise and to avoid litigation (and legislation) the Board agreed to make a payment of $1,500 per year for each child.

There were problems with other governmental bodies in addition to the Council. The University of Oregon planned in 1949 to build its own heating and generating unit. The adverse reaction of the Board was based on the fact that it had ordered a new substation costing $50,000 to serve the university. After the university completed construction of its plant, it sought an interchange agreement with the utility similar to those between the Board and the lumber firms which generated some surplus electricity. The first impulse of the Board was to refuse this arrangement, but in November, 1952 it approved an interchange agreement.

The Board was generous during the post-war period in its participation with other units of local government in long-range planning. In May, 1945 the Central Lane County Planning Council requested the participation of the Board in a study of governmental needs in the area. An initial contribution of $2,500 was made provided that other local agencies made contributions and that the Board be represented on the steering committee. Subsequent contributions in similar amounts were made and a number of long-range plans of general benefit were developed.

Proposals for new construction on the McKenzie River put the Board into greater contact with state agencies during this period. Even before plans were made for development of the upper-McKenzie, the Oregon Fish Commission complained of low water in portions of the McKenzie due to diversion of water into the two power canals at Leaburg and Walterville. The providing of fish rearing facilities at Leaburg Dam at a nominal cost in 1947 eased the situation. Additional consideration of the concerns of the State Game Commission was maintained in the planning for the Beaver Marsh project and an agreement signed with that body in April, 1953 reduced to some extent the number of those opposed to the construction.

After the voters of Eugene declined to authorize Beaver Marsh in March, 1956, the Board had to contend with yet another state agency in attempting to find some means of developing hydroelectric power on the upper-McKenzie. The Oregon State Water Resources Board, established by recent legislation, attached little importance to the generation of power and an inspection trip along the McKenzie River in October, 1956 with representatives of that agency proved disheartening to representatives of the Board. In order to achieve some change in the viewpoint of the Water Resources Board, the Eugene utility joined others in sending a statement to the state agency, presenting arguments for greater consideration of the importance of the waters of the state in the production of power.

June 16, 1955; Beaver Marsh pond area - hydroelectric project rejected by voters. (EWEB files)

LOBBYIST EMPLOYED

The increased importance of state legislation to Eugene's municipal utility eventually led to the employment of a representative to work against undesirable legislation. Although there had been bills before the Legislature of interest to it previously, tax proposals in the early 1950s were of great significance to the Board. In January, 1953 when it was certain that various bills would be introduced which would affect the Board's position or policy, the legal representative of the granges to the Legislature, Mr. Ray Kell, was also approved as the Board's representative.

Searching for additional revenues, Springfield School District No. 19 asked the 1953 Legislature for a law permitting the taxation of municipally owned electric utility properties which lay outside the boundaries of the respective city. Due in part to the activities of the legislative representative of the Board, the measure did not pass. After the Legislature adjourned, the Board authorized that a study be made of the tax problems created by the extension of its services into areas which had been previously served by privately owned-taxpaying utilities. The Board had committed itself to members and committees of the Legislature to do so, but came to the conclusion in September, 1954, following the study, that it should not pay taxes or payments in lieu of taxes to governmental units outside of Eugene. In reaching this conclusion the Board stated that it was not a taxpayer and that its low-cost electricity as compared to charges imposed by investor-owned utilities amply repaid the people living in the area in question. A statement was submitted to the 1955 Legislature by the Board and that session brought forth no adverse legislation, although an interim committee was set up to study taxes including those paid by public utilities. In January, 1956 arrangements were made to have the Board represented before meetings of this committee. Consequently, the committee, while recognizing that there were problems involved in the existing situation, did not see fit to make any recommendations for a change.

Immediately after World War II the Board received much of value from the federal government. In 1946 a bulldozer, pavement breakers, steel plates and a concrete vibrator were acquired from the War Assets Administration at low prices. The next year a crawler crane was secured from the same agency for between 25 percent and 33 percent of its actual market value.

The problems of the Eugene utility with the United States government in the years following the war centered principally around proposed construction of power plants on the McKenzie River. Plans were developed by the Corps of Engineers for multi-purpose dams on the river which had previously been considered the province of the Board. When the Board attempted to become a power-producing partner in these projects, it was thwarted. Although it had prepared a bill providing for a partnership arrangement and presented its case at a hearing held in Washington, D.C. in April, 1954, it ran into objections from at least one member of the Oregon congressionial delegation. Oregon's junior senator, Wayne Morse, was opposed to the principle of partnership in federal projects. In order to strengthen its case the Board joined with Pacific Power & Light Company during 1955 in submitting to Congress a joint partnership bill. This legislation would have involved PP&L in the federal Green Peter Project and the Eugene utility in the Cougar Dam Project. Although some support was mustered—principally from the Eisenhower administration which supported the principle of private partnership in federal projects—the opposition of much of the congressional delegation from the Pacific Northwest, the Hell's Canyon Association, the Northwest Public Power Association and others turned out to be too great and the measure was defeated.

Beaver Marsh, as well as the Cougar Dam project, met with federal opposition. The Regional Forester was opposed to Beaver Marsh in 1954 because of what he thought would be a detrimental effect on the upper and middle falls of the high-McKenzie River. Although the Federal Power Commission was not opposed to construction of the project, hearings which it held in Eugene during June, 1955 provided a forum for opponents. In August of that year the Commission granted intervention rights to the Save the McKenzie River Association, thereby strengthening the hand of a vocal group in opposition to the construction of Beaver Marsh. By March, 1956 Beaver Marsh seemed doomed to failure.

PERSONNEL MANAGEMENT IMPROVES

During Ray Boals' administration a number of changes were made in the personnel policies and programs of the Board. New positions were created, organization charts drafted and the duties of Board personnel more clearly defined. Positions were classified and appropriate pay rates assigned. Increases were made in the general level of wages to compensate for the increasing cost of living and to bring wages paid by the Board in line with those paid by others. Hours of work, overtime, leave and vacation policies were liberalized. New training programs were adopted and employee services expanded. Insurance coverage for employees was increased and retirement benefits instituted. Relations with an internal organization of employees were strengthened and it was necessary to deal with an attempt to affiliate certain employees with an international union.

Following studies by the Public Administration Service, a consulting firm, in the years following World War II, the Eugene Water & Electric Board adopted charts showing relationships among existing positions in the organization and important new positions were created. The consulting firm had first approached the Board in 1946 to make a survey of the personnel situation and after being engaged, made studies in that year and the years following. In March, 1951 Boals presented organization charts to the Board for its consideration and on February 11, 1952 the Board reviewed and approved the organization chart and definitions of duties of personnel. In December, 1954 an important change in the top echelon of the administrative staff occurred after another study by the Public Administration Service. The Board approved the establishment of three deputy directorships—for administration, operations and finance. Also recommended was the addition of a person with training and experience in personnel and administrative work.

The Public Administration Service made a number of recommendations concerning position classifications and the associated pay rates during this period. On the day that General Superintendent-Secretary McArthur tendered his resignation, January, 14, 1946, he also called the Board's attention to the proposed survey and salary study by the consulting group. The matter was left to the discretion of the incoming general superintendent-secretary, Ray Boals, who decided to proceed with the project. The study was concluded and the Board adopted its recommendations as to job classifications and pay ranges on June 18, 1946, with pay increases retroactive to April 15. Some modifications were made in the plan after employees protested, but on the whole it was quite satisfactory, providing a reasonable means of wage administration. Additional studies were made by the Public Administration Service in 1952 and 1954 with its recommendations on wage and job classifications adopted by the Board. By 1956, after a personnel officer had been hired, the Board was in a position to make its own studies and arrangements with the Public Administration Service were discontinued.

In the post-war period changes were made in the general level of wages as well as in relationships among wages paid for different positions. In January, 1947 petitions were received from a number of employees in which they asked for a cost-of-living adjustment. The Board agreed and after rapidly increasing prices brought another such request later in the year, a policy was adopted of making general wage adjustments on the basis of the Consumer Price Index published by the Bureau of Labor Statistics.

Working hours were shortened and overtime provisions liberalized between 1946 and 1957. Because of confusion and dissatisfaction with the existing system (under which overtime was paid only after 44 hours were worked during any week), the Board authorized in November, 1946 the payment of time and a half for all work on Saturday afternoons and Sundays regardless of the number of hours worked during the rest of the week. Also, holidays were to be counted as time worked for the purpose of computing overtime pay. The Board adopted the 40-hour week beginning August 1, 1950 with payment of overtime for hours worked beyond 40 hours to begin September 1. On September 10, 1951 the Board decided to apply the rules and regulations regarding pay for overtime as set forth in the Federal Wage and Hour Law to monthly salaried employees. In August, 1952 the Board approved an 8:00 a.m.-4:30 p.m. day for office employees with one-half hour for lunch. Finally, the Eugene utility changed the overtime rate to double time in November, 1956 to bring its practice into accord with that of other utilities.

Various leave policies were also changed in the post-war years. Early in 1951 a committee of employees asked for increased vacation time. No action was taken until April, 1954 when the Board adopted a vacation schedule of 15 working days for all employees with 15 or more years of service, with other employees continuing to receive one hour of vacation for every 26 hours of employment, or 10 working days per year. During May, 1950 the Board approved the granting by the general superintendent-secretary of leave on a partial salary to two members of the Naval Reserve for a two-week training period (this in addition to their regular vacation). At the same time, when the question of vacation was raised in 1951, the employees also asked for a more liberal sick leave policy. The Board on February 12, 1951 acted favorably on this request when it resolved to permit cumulative sick leave to increase from 30 to 60 days. It further resolved to pay the difference between the amount paid by the State Industrial Accident Commission and the employee's full salary during periods of work-related disability. Further modifications were made in sick leave policy in 1955 when recommended by Deputy Director-Administration Henry Beistel.

To improve the effectiveness of its employees, the Board established policies providing off-the-job training in the middle 1950s. Ten operators were sent to a waterworks operators short course at Oregon State University in March, 1954 with regular wages paid them while in attendance. In November, 1955 the governing body approved a general policy on training courses for employees under which it would provide one-half the cost in certain cases and also provide time off with pay plus expenses for approved attendance at conferences.

Employee services were also expanded during Boals' administration. The Board paid the entrance fee for an employee softball team in the city league with the employees to sell season tickets and return the proceeds to the Board. After the new office and warehouse were built, kitchen and dining equipment for the assembly room was furnished and hot lunches provided at nominal prices. Employee service pins were authorized in January, 1953 to give recognition to long-service employees.

INSURANCE PLANS EXPANDED

Group insurance plans were liberalized in the post-war years. In September, 1947 the Board permitted employees to continue their group life insurance after retirement with the employee paying the entire premium. The city of Eugene adopted a policy in January, 1955 of paying one-half of the premiums on group life insurance up to $1,500 for its retired employees and in February, 1955, the Board did the same. Although the matter of sickness and accident insurance had been discussed again in 1946, it was not until 1956 that the Board approved a comprehensive health and hospitalization plan. This covered active employees in addition to those retired employees who had been employed 10 years or more. The Board paid the premiums for employees and retirees only with both groups paying for dependent coverage if desired.

Retirement benefits were also expanded. Before 1946 the Board on occasion gave a sum of money to a retiring employee, but there was no retirement plan in effect. After the state had passed legislation creating a Public Employes Retirement System, the Board, after some delay, decided on March 11, 1946 that it would adopt the plan. Five years later, on June 11, 1951, the Board went on record as favoring an increase in retirement benefits as provided in new state legislation after being advised that the City Council and Park Board had expressed themselves as being in favor of such improvement in benefits. At the end of 1956 other changes were made in retirement policy, with the retirement age set at 65 and other provisions adopted in accordance with those of the Public Employes Retirement System.

Throughout the period 1946-1957 the Board and its administrative staff had dealings with a committee of employees. This group grew out of a safety committee existing before the war and provided a means for communicating the general views of the employees to the Board and its staff. After an outside union had talked of organizing the employees of the Board, the employees' committee was organized on a more formal basis in 1950 to meet with the Board for a general discussion on wage policies. In February, 1956 this committee appeared before the Board in order to point out the need for more attention to the personnel problems of the organization. One specific comment was that the rules and regulations governing the employees and their activities had not been brought up to date for some time and that there was a definite need for the establishment of rules and personnel policy. It was a further feeling of the committee that no attention was being given to personnel management. The Board replied that it would give general approval to the hiring of a personnel officer, but that personnel matters were the responsibility of the administration and that the committee should take them up there, coming to the Board only if they could not be dealt with by administrative personnel.

Beginning in 1949 the Board had some serious problems with organized labor groups. Local labor unions were concerned with the fact that some non-union workers were being employed on the construction projects for which the Board had contracted. More importantly, it began to appear that an attempt would be made to organize the Board's employees. In October, 1949 the Board's attorney advised that employees had the right to join labor organizations and to picket, but that the Board was not required by law to pursue collective bargaining with any labor organization. The Board was not approached by any union representative until September, 1950, when an officer of the International Brotherhood of Electrical Workers appeared at a Board meeting to discuss the organization of EWEB employees. He proposed to organize the workers in the electric department under the Electrical Workers Local and those in the water department under the Plumbers Local and asked the Board's permission to discuss the matter of organization with the employees. The Board stated that the matter was one which the employees themselves would have to decide and nothing more was heard of this for a time.

Union officials again confronted the Board in 1956. In August of that year the Board received information that at least 60 of its employees had joined the International Brotherhood of Electrical Workers. On October 15 representatives of the union, including several employees, presented to the Board a statement of objectives of EWEB employees who had become members of the union, stating that 101 employees had signed up, and requested that the Board consider a formal agreement with them. This request was referred to the Board's attorney and staff for study. In December the opinion of the legal counsel was presented to the IBEW members and the matter was again dropped. The opinion reaffirmed the earlier position that the Board was not legally compelled to enter into collective bargaining.

During Boals' administration there were many accomplishments. The construction program he had proposed when he took over had been carried out with some additional projects as well. Water and electric service had been extended and improved with electric rates brought to a new low level. There still remained the question of how to get public approval for the construction of additional hydroelectric plants on the McKenzie and its tributaries.

In March, 1957 Ray B. Boals—well past retirement age—tendered his resignation as general superintendent-secretary, remaining on the staff as a consultant to the Board.

Ray Boals was a warm and generous human being. His sense of humor was profound and his technical competence admired. He was fond of chiding in a good-natured way new employees about their "fire insurance." This was his way of making mild inquiry into the church affiliation of new personnel. In an employee survey conducted near the end of his administration, one comment made was that in order to succeed at EWEB, as the organization had become known, it was necessary to be a graduate of Oregon State University, a Mason and a member of the Central Presbyterian Church. This comment, perhaps offered in whimsy, in no way reflected the personality and character of Ray Boals. He had a deep devotion to all of the work force and was a true gentleman in every sense of the word.

The Board was now confronted with problems still facing it and those yet to come. Board members looked for a new "general" and found him in Byron Price.

CHAPTER SIX

Byron Price—A New Era

WHEN BYRON PRICE JOINED THE Eugene Water & Electric Board, 46 years had passed since the original Eugene Water Board held its first meeting in 1911. The municipal utility operation had demonstrated steady growth over the years in both the water and electric systems. During 1956 the average number of water customers was 11,130 and electric customers averaged 28,159. Electric revenues reached $3.5 million and water revenues were $621,378 for the period. The Eugene system had established itself as one of the principal municipally owned utilities in the Pacific Northwest. As compared to the larger municipal systems in Tacoma and Seattle, the Eugene Water & Electric Board was unique in that its elected Board maintained a greater degree of autonomy with respect to city hall than was and is the case with the larger systems to the north.

Price's immediate past employment was as assistant administrator for the Bonneville Power Administration. A native of Montana, he attended Montana State University, majoring in law. His first employment in the utility field was with the Montana Water Conservation Board in Helena. In 1941 he joined the Army Corps of Engineers in which he served both as a military officer and then as a civilian until 1953. After leaving the Corps, Price joined the Eugene utility system for one year as administrative assistant to Superintendent-Secretary Boals. He then joined the Bonneville Power Administration from 1954 until July 1, 1957—the date on which he became the new general superintendent-secretary for the Board.

With the advent of the Price era there were no fundamental changes in the objectives of the Board. However, Price proposed changes in policies and programs by which those goals could be reached. In his early discussions with the Board he set forth a five-point plan for addressing the problems still confronting the utility. First, he proposed a program to secure public support for building much-needed generating capacity on the upper-McKenzie. Second, he stated his intentions of entering into more cooperative relationships with other electric utilities in the Pacific Northwest. Third, he proposed a comprehensive program for improving community relations. Fourth, he would seek means for securing improved relations with the City Council, school districts and other governmental units. Last, but certainly not least, Superintendent-Secretary Price made proposals for bettering employee relations while improving organizational efficiency.

Price, who was age 42 at the time of his appointment, did not have the educational credentials possessed by his three predecessors, McClain, McArthur and Boals. He was not an engineer and had not received a college degree. However, he had characteristics which more than compensated for these deficiencies. Not only was he indefatigable, he also possessed an extremely high intellect which was to serve him well. His top staff as he took charge of the utility operation was of the highest quality. Byron K. Taylor, age 53, was deputy director-operations. He was a civil engineering graduate of Oregon State University and had joined the utility operation in 1928. Ralph B. Patterson, 57, deputy director-finance & accounting, was first employed by the Board in 1933. Henry F. Beistel, deputy director-administration, was 50 years of age. Although he had only been with the utility operation for about two years, Beistel had many years of experience with the University of Oregon's Bureau of Municipal Research and with the city of Eugene as finance officer. He was a graduate of the University of Oregon. Harold H. King, controller, was 52 years of age. Also a graduate of the University of Oregon, he had first worked for the utility in 1926 on a part-time basis. King left the organization to enter private accounting practice for a period and returned in 1946 to become the most knowledgeable employee in the areas of statistics and accounting.

Vernon J. Poindexter, age 54, was electric operations superintendent. He had been with the utility since 1928 and is credited as being a driving force in the construction of the electric system as it existed in 1957. Kenneth W. Rinard, age 30, was water operations superintendent. A graduate of Oregon State University, Rinard was considered a brilliant mechanical engineer. The position of chief engineer was held by Ralph Martin, Jr., age 36, who had joined the Board in 1949. He had been chief engineer since 1954.

There were about 265 employees during 1957. This work force, in addition to those listed above, included many individuals who made significant contribution to the Board and its purposes.

When Byron Price joined the organization an anomaly existed in the system of titles. As the reader might already have ascertained, there were three deputy directors, implying that there was someone who held the title of director. This kind of confusion is common to many organizations, both in the private and public sectors. Price was aware of this and was eventually to correct this nomenclature. But for the time being more pressing issues were at hand.

Byron Price; general manager-secretary of the Eugene Water and Electric Board 1957-1973. (EWEB files)

QUEST FOR ADDITIONAL GENERATION RENEWED

The major problem confronting the Eugene utility and Price after he became its chief administrator was the question of how to develop additional power so urgently needed. A report on the Beaver Marsh project—which had been turned down by the voters in 1956—indicated that it was still a feasible project with many advantages. However, on September 9, 1957 Price recommended that Beaver Marsh be held in abeyance until the Carmen-Smith project could be investigated. It appeared that the Save the McKenzie Association would not object to the building of the Carmen-Smith project. In October, 1957 Price was ready to move ahead with planning for new hydroelectric generation. He reviewed for the Board the original Beaver Marsh project, mentioned studies of possible sources of hydro power on the Siuslaw and Umpqua Rivers and also discussed the present and future power available to Eugene from Columbia River projects. His conclusion and recommendation was that the Board go ahead with planning the Carmen-Smith project and that the State Water Resources Board should be asked to hold a public hearing on hydroelectric development of the upper-McKenzie with Carmen-Smith as an alternative to Beaver Marsh. The Board approved hydrological, topographical and geographical studies, an analysis of downstream benefits and other research that would lead to successful license application.

In January, 1958 Superintendent-Secretary Price informed the Board that Senator Richard L. Neuberger and Congressman Charles O. Porter were favorable toward Carmen-Smith. Neuberger, along with Senator Wayne L. Morse, had opposed Beaver Marsh. In February, Burke Hayes of the engineering firm of Cornell, Howland, Hayes, and Merryfield (CH_2M) presented a Carmen-Smith plan to the Board and it was decided to seek voter approval of the project. The engineering firm's report estimated that energy available from this project would cost less than that purchased from BPA and that the project seemed to be the most economic and feasible way of securing additional electrical energy. Upon receiving these recommendations supporting the conclusions of Byron Price and his staff, the Board on February 24 authorized the filing of applications for construction of the Carmen-Smith project with the State Engineer and the Federal Power Commission. Proceedings for bringing the matter before the voters were begun.

Governmental approval of Carmen-Smith was expeditiously secured. On September 22, 1958 the Board was notified that the State Water Resources Board and the State Engineer had approved the proposal and on January 8, 1959 the Federal Power Commission issued an order granting a license for construction.

After receipt of the Federal Power Commission license the Board was ready to seek voter authorization. On January 21, 1959 a resolution was sent to the City Council asking that it submit—at a special election—a charter amendment to the voters authorizing construction of the project and the issuance of revenue bonds to finance it. Originally, the amount and other terms of the revenue bond issue proposal were to be left unspecified to permit greater flexibility in constructing and financing the project. However, the amount was later established at $25.5 million. Importantly, the bonds were not to be a general obligation of the municipality, but rather were to be redeemed totally out of the revenues of the electric utility.

Many steps were taken to secure citizen support of the Carmen-Smith project and the issuance of bonds to finance it. A statement explaining the need for the facility and its financing was drafted for the voter's pamphlet. The Board also agreed to prepare a notice of the special election and its importance to be mailed to all customers within the city limits prior to the election. A 30-minute motion picture was produced by the staff explaining the project which was shown widely in the community and on television.

If a single factor can be judged most important in the campaign prior to the election, that factor would be Byron Price himself. Price personally attended meetings of many groups and mollified what might have become hostile reactions toward the project. His skills in this kind of activity excelled and he was known to refer to himself from time to time as a "political engineer." Later reflecting upon this period, Price said that a two-way process was involved. While he imparted a great deal of information to various groups, he also gained much information that was helpful to the campaign.

The election on April 28, 1959 met with success. The citizens of Eugene approved the amendment by a ratio of almost 3 to 1 (yes, 2,991; no, 1,025). A proclamation by the city recorder dated May 11, 1959 put the Board in the position to proceed legally with the construction of Carmen-Smith and to sell the necessary bonds.

After voter approval of the bond issue an unforeseen financial problem arose. Further engineering studies in the fall of 1959 indicated that the most cost-effective installation would be a 94,600 kVa (about 90,000 kw) facility costing $28.45 million rather than the originally planned construction of a 56,250 kVa facility costing $23.5 million. If the larger capacity were to be installed, this would pose a financial problem in paying interest and principal on the higher amount while at the same time securing funds through retained revenues for other necessary capital improvements involved in electric system expansion. The alternatives considered by the Board were: (1) selling surplus power after the project was completed; (2) reducing the scheduled capital improvements; (3) increasing electric rates; and (4) building the project in two stages. On November 30, 1959 the Board approved the construction of the larger project and authorized Superintendent-Secretary Price to enter into negotiations to sell the surplus power.

Construction of the Carmen Smith project was begun during the summer of 1960. On July 19 the Board approved selling the entire $25.5 million of bonds to Blythe & Co., Inc., Halsey, Stuart & Co., Inc. Associates. Construction bids had been opened on July 8, 1960 and Peter Kiewitt & Sons' Company was low bidder. The staff was authorized to issue a letter of intent to the Kiewitt firm. Bechtel Corporation was selected as project engineer.

Also in 1960 the Board gave its attention to the water system and the need for additional construction. Contracts were awarded on May 9 to build the one-million gallon City View Reservoir and in June, Deputy Director-Operations Byron K. Taylor advised the Board of the need to plan for a new reservoir in the Hawkins Hill area. The Board approved preliminary plans for this project and then heard a detailed report from Chief Engineer Ralph Martin, Jr. on future water system needs. Martin, in his ten year projection, spoke of the need for determining all possible water sources including wells and the headwaters of the McKenzie River and the economic aspects of such sources. He said that future water needs should be considered on an area-wide basis rather than only Eugene and that the legal, engineering, economic and social aspects of the quantity and quality of water for both domestic and industrial use should be considered. It was the consensus of the Board and staff that the development of water resources should be on a metropolitan basis. The issue was pursued further in December, 1960 and the Board authorized application to the State Engineer for additional water from the McKenzie River.

DIALOGUE WITH CITY CONTINUES

The dialogue between the Board and the City Council continued with topics of discussion much as they had been in early times. The two bodies met in September, 1959 to seek resolution of their disagreements. Discussed were charges for services made by each party against the other, payments made by the Board in lieu of taxes and a proposal to improve street lighting in the downtown area. An immediate result was the decision by the Council to modernize the street lights. Less tangible, but perhaps more important, was increased understanding between the two bodies. Additional informal meetings were held and the consequent improvement in understanding of the mutual and respective problems of the Council and the Board led to increased cooperation and coordination of activities.

CONTRIBUTION TO LOCAL GOVERNMENT EXPANDED

Through action of the 1957 Legislature, municipal utilities were permitted, not mandated, to make in-lieu-of-tax payments to school districts. This was a concern of the Board and after an investigation of the history of such payments, Superintendent-Secretary Price recommended that the Board adopt a plan wherein a determination would be made of the amount that a privately owned utility would pay in local taxes. He proposed that after the contribution of six percent of gross electric revenues to the city of Eugene was deducted, the balance available be divided up among the school districts in which the utility sold electricity. Price explained that neither an ad valorem or sales tax approach was workable or applicable and that it would be necessary to use methodologies employed by the Oregon State Tax Commission in order to arrive at a proper solution. On the basis of Price's analysis the Board approved the following distribution of payments for 1959-60: City of Eugene - $228,600 (six percent of gross electric revenues); School District 4 - $32,000; School District 52 - $14,000; School District 19 - $28,000; and School District 68 (McKenzie) - $5,000. This schedule of payment proved satisfactory to the recipient school districts. A special problem arose in September, 1960 when McKenzie School District 68 in the Blue River area provided education to the children of engineering and construction personnel who were living and working on non-taxable facilities—federal public land—on the Carmen-Smith project. The issue was resolved by an agreement by the Board to buy a school bus for the children and to pay $30 per month per student to the school district.

The annexation of part of the service area of EWEB to the city of Springfield placed the Board in the position of taxpayer to the sister city. Some 1,000 electric customers of the Eugene utility were in the territory annexed, and a payment of three percent of gross electric revenue received from these customers was made in lieu of property taxes commencing in July, 1960.

The Board, as it existed in 1957, proved to be a powerful ally for Byron Price and the programs he undertook. The Board was solidly representative of the business community. Lionel W. Trommlitz, a Board member since 1944 and a retired radio station executive, served as president. Oluf A. Houglum, a retired banking executive and a Board member since 1951, served as vice president. Calvin R. Smith, manager of the J.C. Penney store in Eugene, had been appointed in February, 1954. Earl L. McNutt, who had served the community as mayor, state legislator, president of the Chamber of Commerce, member of the Bonneville Advisory Committee, and as a member of three Lane County boards, had been a member of the utility board since 1954. Julio W. Silva, who owned the Chevrolet dealership in Eugene, joined the Board in 1956. These Board members met the needs of the time. They were supportive of Price and were interested in the growth of the utility operation and the community. Oluf Houglum is credited with guiding the utility toward the sale of revenue bonds instead of traditional general obligation bonds.

Relationships with a variety of diverse factions were sharply enhanced under Price's administration. In 1959 Price became chairman of the Pacific Northwest Utilities Conference Committee, a group composed of both consumer- and investor-owned generating utilities. Harold H. King, controller, became treasurer of that organization at the same time. The PNUCC, as it was called, represented a vehicle through which utilities of the Northwest could work cooperatively together toward serving their interests and those of their customers. This organization worked to support the plans of the Bonneville Power Administration, the Army Corps of Engineers and the Bureau of Reclamation for additional generating facilities in the region which would provide EWEB and other utilities new sources of low-cost power. Much was gained through the working relationships which were established and in 1960 arrangements were made with Pacific Power & Light Company to store excess energy generated at the Priest Rapids Project in reservoirs available to PP&L through an inter-company pool. Thus the Board was able to avoid spilling its share of energy from that project when it was in excess of its own requirements. Byron Price played a key role in the activities of the Pacific Northwest Utilities Conference Committee, and on several occasions journeyed to Washington, D.C. to represent this organization on regional affairs.

Originally known as the Tacoma Conference, the Pacific Northwest Utilities Conference Committee was to play an extremely important role in regional affairs and Byron Price provided leadership in this process. The Columbia River Treaty was ratified in 1964. It was designed to provide benefits of hydro generation and flood control to both Canada and the United States and was the result of the work of many individuals and organizations including PNUCC. Through coordination agreements, the Columbia Storage Power Exchange and innovative financing in the purchase of Canadian entitlement to power under the Columbia River Treaty, the region has gained vast benefit on both the American and Canadian sides of the border. Price served EWEB and the region well in these endeavors.

The relationship with the customers of the utility was becoming more complex. The Board was still in a promotion mode with respect to the sale of electricity. Programs for adequate wiring and both gold and bronze medallion homes were in place. As early as 1934 an all-electric rate was instituted under Superintendent-Secretary McArthur. In the early 1950s residential demand meters were installed. This caused a great amount of confusion and generally there was customer discontent about existing electric rate structures. Superintendent-Secretary Price recommended in November, 1957 that an independent engineering consulting firm review the rate schedules and practices. No affirmative action was taken by the Board until October, 1959 when a full-scale study of electric rates, rules and regulations was authorized. The policy of imposing a 10 percent surcharge on electric customers outside the city was to be included in the study, as were the electric rules and regulations of EWEB as compared to the practices of other utilities in the area. After the consulting firm of Ford, Bacon & Davis made its report to the Board in September, 1960, a new residential electric rate schedule eliminating the demand feature was put into effect on October 1, 1960 with the 10 percent surcharge to customers outside the city remaining in effect. The new rate provided the first 20 kilowatt-hours at three cents, the next 80 at 1.5 cents, and all in excess at 0.8 cents.

SURCHARGE ELIMINATED

The 10 percent surcharge was to continue until March 11, 1963 when it was eliminated. The Board apparently felt that it was unfair to discriminate against out-of-town customers who, through the basic rate, were contributing as much on an individual basis to the utility as were in-town customers. The fact that the electric utility had begun using its total system electric revenues rather than the credit of the city as the basis for selling bonds also contributed to this decision.

In September, 1958 the Board reaffirmed its policy of not providing customer lists to anyone. The Board did, however, continue its participation in community affairs in certain ways. On March 9, 1959 the Board acted to contribute $1,000 for the Lane County Centennial exhibit. At that same meeting it voted to participate in the activities of the Electric League of Lane County through contributing 10 cents per electric meter per year and the services of a half-time employee. This organization, which was to become the Electric Association of Lane County, had as its principal members EWEB, Springfield Municipal Utility Board, the two REA cooperatives, contractors, architects, wholesalers and retail dealers.

Price also sought greater support of the chambers of commerce in the area. Already a member of the Eugene and McKenzie River chambers, Price recommended in November, 1960—and the Board approved—membership in the Springfield Chamber of Commerce. This action was taken after the city of Springfield annexed territory which was being served by EWEB. Dues to the Eugene Chamber were increased from $1,800 to $3,000 early in 1960. This was preceded by a change in the policy regarding participation in installing Christmas decorations in the downtown area. Prior to 1958 the Board charged for such special services, but beginning with the Christmas season of that year, authorization was given for the use of EWEB equipment and employees in the installation and removal of street decorations after a Chamber of Commerce committee made such a request.

Water Board Park had been established in 1955. This facility became and remains an important recreation asset of the community. This kind of concern for the community and the efforts by Byron Price to enhance community relations at all levels brought results. As early as December,

1957 when a hearing was conducted by the State Water Resources Board on the Carmen-Smith project, there was little or no intense emotion displayed by objectors to the proposed construction. EWEB was complimented during the hearings for its efforts toward the protection of recreational and scenic values in its plans for power developments on the upper-McKenzie. The objectors were limited primarily to those who categorically objected to any such development. Another goodwill measure was achieved in 1958 when the Board arranged with the Lane County Fair Board to build and maintain a high-voltage service system for the fairgrounds since the Fair Board did not have experienced men for such work.

The Board was always on the alert for additional power supplies. During 1958 the Board approved purchasing, under contract, two percent of the power from the Wanapum project to be built by the Grant County Public Utility District. The arrangement was similar to that through which the Board purchased power from the Washington utility's Priest Rapids project. The cost of the power was expected to be 3.7 to 3.9 mills per kilowatt-hour, slightly more than the cost of Priest Rapids power. Although Wanapum power would not be available until 1962, the Board was looking ahead to meeting future power needs.

A pollution problem above its intakes on the McKenzie River forced emergency action on the part of the Board in 1958 and 1959. After receiving advice and recommendations from county and state health authorities, increased chlorination of the water was authorized in December, 1958 to combat a special pollution condition upstream from the Hayden Bridge Filter Plant intake. To meet the situation more positively, arrangements were made with the Weyerhaeuser Timber Company's Springfield plant to divert the problem "runoff" into its wastewater discharge line. In studying a more permanent solution, it was proposed that water from the tailrace of the Walterville Plant be transmitted through a pipeline or canal directly to the filter plant where it might generate power before being filtered. This plan was never implemented.

Rapid growth in the number of EWEB water customers and increases in individual customer consumption meant that Eugene was outgrowing its existing water system by 1960. The Hayden Bridge Filtration Plant, only about a decade old at that time, could be expanded but only at what was expected to be considerable cost. However, EWEB's Winston Berkeley, assistant water operations superintendent, was intent upon finding a method which would improve the plant's capacity and finished water quality with a minimum of capital outlay.

TREATMENT CAPACITY INCREASED

Berkeley recommended an innovative water treatment method. First introduced at the Atomic Energy Commission's water treatment plant at Hanford, Washington, the Pit-Con water treatment process was developed by two General Electric engineers, Ray Pittman and Walter Conley. The Pit-Con process permitted filtration at rates from two to five times faster than was typically possible with normal rapid-sand filters. This process, which utilized carefully graded anthracite coal as a replacement for sand in the upper sixteen inches of the Pit-Con filters, would yield a capacity at the plant of an additional 27 million gallons of water per day. The cost of conversion to the new process was about $50,000. This was in contrast to the $700,000 which would have been required to expand the plant to a comparable capacity using conventional rapid-sand filters.

The Pit-Con process was a revolutionary development which gained world-wide attention. Win Berkeley, without formal credentials, was nevertheless respected as a water-treatment scientist. He was admired for his professorial manner and thoughtful approach to problems. Following retirement, he performed extensive work as a consultant in water treatment.

Relations with the federal government became more important. During the Carmen-Smith application period, although some trips to Washington were made by Board personnel, a Washington attorney was employed to maintain continued contacts with the Federal Power Commission and the Department of the Interior. Since the Carmen-Smith project was to be located in the Willamette National Forest, agreements with the Forest Service were also necesary, and final approval was given by the Board to arrangements with that agency in September, 1959.

INTERTIE OPPOSED

Proposed federal legislation drew the attention of the Board in 1960. In January it went on record opposing the proposed BPA intertie with California utilities. In June the Board supported an amendment to the Federal Power Act which would provide payments to those utilities providing water control or storage facilities. EWEB was concerned primarily because of its participation in the Priest Rapids and Wanapum projects in Washington.

Another "public" of vital concern to Byron Price was the work force. As a token of its interest in better employee relations, the Board in December, 1957 approved a recommendation of the superintendent that the Christmas gift to employees be increased from $10 to $15. Communication with employees was enhanced and a major step taken in 1958 when comprehensive personnel rules and regulations were adopted. That same year the Board raised wages and salaries at all levels in the organization, followed by additional increases in 1959.

In November, 1958 a comprehensive employee training program was adopted by the Board. This policy provided generous financial participation on the part of the Board and has been proven effective over the years. The policy has paid dividends not only to participating employees, but to the total organization.

The new personnel rules and regulations were effective January 1, 1958. They were prepared from various memoranda, verbal instructions and informal practices and were also based to a large extent upon prevailing practices in Northwest utilities. Greater consideration was given to suggestions received from employees. It appeared that much of the personnel discontent of the past was the result of lack of written rules and regulations and the new administrator felt that this was a big step forward in solving personnel problems. Noteworthy was Price's belief ". . . that such rules and regulations should be the responsibility of the superintendent . . ." The appeals procedure included in the rules and regulations provided that the decision of the superintendent would be final. This was in keeping with Price's management philosophy that administration, including personnel matters, was the province of the superintendent and not of the Board.

At the same time the new rules went into effect, improvements were made in pay. The linemen and wiremen received an increase after the annual wage survey was made based upon patterns found among several Northwest utilities. A new form of pay plan for hourly employees

March 17, 1959; supervisory training. (Left to right) Laudie Kralicek, Les Pearson, Ben Barnts, Darwin Spricer, Ross McClanahan, John Hauck, Victor Blunt, Nathan Chaffee, Bobby Chalcroft, Leonard Boylan, Harold Buell, Sam Pakulak, Orval Dyer, Earl Chapman and Prof. Saville T. Ford, University of Oregon. (EWEB files, Bob Miller photo)

was adopted, also following the practice of other utilities. Although no change was made in the classification and salary of most of the monthly employees, adjustments were made in the accounting and engineering departments to bring salaries there into comparative position with those paid in other utilities and local government. Administrative and supervisory salaries were also advanced based on the duties and responsibilities of the positions involved.

Salary adjustments were also made in 1959. The annual survey indicated that EWEB salaries were a little higher than city, county and other local salaries, but a few cents an hour below the other electric utilities in the Northwest. In January the Board approved a two percent cost-of-living increase for all employees, but deferred consideration of other wage and salary increases until after other utilities had concluded their negotiations with unions. Agreements were reached in other parts of the industry in April, and effective May 1 the Board raised linemen's wages to the level contained in the union contracts and gave an across-the-board increase of 3.2% to all other employees.

Changes were also made in vacation, sick leave and retirement policies after Price took office as chief administrator. Long-service employees received a vacation of four weeks after 20 or more years of service, beginning in 1958. In January, 1959 a new leave policy was established, changing the existing method of administering sick leave to a more desirable means and also permitting employees to accumulate 30 days of annual leave. Finally, in July, 1959 the Board agreed, after investigation, to make greater contributions for certain employees under revised Oregon Public Employes Retirement System legislation.

Byron Price undertook the solving of a special personnel problem relating to compensation. There was considerable justification to the feeling that the water department had been treated for a number of years as a "poor relative" in terms of pay levels. An obvious morale problem existed. Price undertook to bring pay levels in the water department into a more reasonable relationship with electric department pay rates and it can be said that the superintendent-secretary did a great deal toward improving the morale of water department employees.

Byron Price also maintained a strong interest in industrial safety. The first safety manager was hired in December, 1957 and a recognized, formal safety program was underway. In an industry with inherent hazards Price saw to it that modern, man-lift equipment was added to the operation. He also encouraged safety meetings and training.

As the 50th anniversary of the Eugene Water & Electric Board approached, Price felt that his initial programs had been satisfactorily carried out. Relations with employees and the customer public had been improved while EWEB had been moving toward its objectives in a highly satisfactory manner. Additional supplies of energy had been secured or were under construction, insuring the availability of low-cost electricity in Eugene for some time to come. Operating efficiency was improved by holding the labor force relatively constant and providing it with better equipment and by automating some of the operations. Although some problems still remained and new ones were inevitable, the Board and administration of the utility looked to the future with justifiable confidence.

CAUTION EXERCISED IN NUCLEAR INVOLVEMENT

The Board exercised caution in considering involvement in the young nuclear power industry. In January, 1960 and again in January, 1961, the Board rejected proposals by the United States Atomic Energy Commission to participate in atomic power reactor demonstration projects. Reasons given for rejecting these proposals in both instances related to the heavy involvement in hydroelectric generation.

The Board continued to give careful consideration to requests for contributions. On March 14, 1960 a request by the science club at South Eugene High School that the Board provide financial assistance to the guidance program of the Junior Engineers and Scientists Institute was denied. However, the Board continued to provide assistance in ways that did not involve expenditures and were a general service to the community. In July, 1960 the Board authorized the addressing of envelopes for the Lane County Tuberculosis and Health Association and subsequently approved many similar requests.

October 12, 1962; John Cochrane (left) and Ralph Patterson manning phones during outage caused by Columbus Day storm. (EWEB files, H. Gross photo)

October 13, 1962; intersection at 11th. Ave. and Patterson—aftermath of storm. (EWEB files, H. Gross photo)

October 13, 1962; University of Oregon campus. (EWEB files, H. Gross photo)

In-lieu-of-tax payments were given close attention by the Board. In January, 1961 payments totaled over $400,000 for 1961-62 and by 1971 the Board would approve in-lieu-of-tax payments in excess of $1 million.

On March 13, 1961 the Board's 50th anniversary was celebrated with a cake in the EWEB cafeteria. Incumbent Board members Lionel Trommlitz, Oluf Houglum, Earl McNutt and J. Orville Lindstrom were in attendance, as were former Board members W. R. Robertson, Percy Brown, Orlando Hollis, E. S. Wengert, Paul Green and Julio Silva. The 1960 Annual Report was designated as an anniversary issue. There was justifiable pride on the part of the Board and the EWEB work force. The first 50 years of the organization represented a period of progress and accomplishment, but challenges remained.

On June 12, 1961 Price reported that Carmen-Smith construction was on schedule. However, many problems were to beset the project before its completion. Labor problems arose, including strikes of carpenters, operators and ironworkers. Power outages and adverse weather conditions resulted in delays and unforeseen problems involving excess water principally in the diversion tunnel caused additional delays and extra expense. On January 8, 1962 the Board approved special incentive pay for the tunnel crews on the project in order to expedite the work. Special crews also were brought from Canada.

Completion of the Carmen-Smith project was originally scheduled for the fall of 1962. However, because of labor strife and other problems, on July 16, 1962 the completion date was changed to February 13, 1963. It was the desire of the Eugene utility to operate the recreational facilities at Carmen-Smith in a manner similar to Water Board Park. However, the United States Forest Service contended that operating such facilities was its responsibility and it eventually prevailed in its contention.

Completion was not realized in February of 1963, and on May 20 the Board was told that cracks had developed in the power tunnel which must be corrected. It was estimated that this work could be accomplished during a period of three weeks working seven days a week.

It was not until June, 1963 that the Board was in receipt of a letter from Bechtel engineer E. W. Peterson stating that EWEB could take over a substantial portion of the Carmen-Smith facilities and proceed with commercial operation. Finally on August 26, 1963 the Board was informed that the Carmen-Smith project would go into commercial operation on August 27 at 12:00 midnight. At the same meeting Henry Beistel reported on plans for the Carmen-Smith dedication which was scheduled for Monday, September 9, 1963. Also at this time Byron Price and Byron Taylor were commended for their contribution to the completion of the project.

The Carmen-Smith dedication was a gala event for Eugene. Some 1,800 citizen attended the dedication at which box lunches were served free of charge. Featured speakers included Governor Mark O. Hatfield;

Eugene Mayor Edwin E. Cone; Orlando J. Hollis, former Board member and dean of the University of Oregon Law School; J. Herbert Stone, Regional Forester, U.S. Forest Service; M. Boyd Austin, Regional Engineer, Federal Power Commission; Calvin R. Smith, vice president of the Board; and F. F. "Monte" Montgomery, state representative. The Rev. Wesley G. Nicholson of the First Congregational Church provided the invocation. Music was provided by the Eugene Summer Music Clinic Band under the direction of Byron Miller.

Over 200 EWEB employees attired in white coveralls and yellow hard hats assisted in the dedication. The activities were spared the thunder and lightning storm which occurred late in the afternoon after the ceremonies had been concluded. It was a proud day for Byron Price, and even though construction had been disrupted by labor problems, adverse weather conditions and unexpected geological phenomena, the Carmen-Smith generating station would become a valuable asset to the Eugene utility and its community.

FLOODS INUNDATE POWERHOUSE

On January 16, 1964 a butterfly valve at the bottom of the penstock entering Carmen Powerhouse failed, flooding the powerhouse and necessitating repair and replacement of the total control system. Again,

EWEB lineman repairing storm damage. (EWEB files, H. Gross photo)

at Christmastime, 1964, major flood conditions resulted in a second inundation of the Carmen powerhouse. Corrective measures were taken to prevent further such occurrences. Notwithstanding those early difficulties experienced with the Carmen-Smith project, the facility has generated an average of about 250 million kilowatt-hours each year, providing a very valuable capacity factor for Eugene's energy needs with attendant benefits to Eugene ratepayers.

On March 27, 1961 the Board approved budget billing for its customers. This process permits customers to levelize their payments over an entire year and also to avoid very high bills in the case of electric heating customers which would occur during cold periods. Budget billing at EWEB was an example of a program conceived by one part of the organization but not well-received by other components. The accounting department was uncomfortable with budget billing and discouraged its promotion. Although there is organization-wide acceptance of budget billing today, at this writing, only 3,054 customers out of some 66,000 have availed themselves of this opportunity.

MAJOR ROLE IN AIR POLLUTION CONTROL

Superintendent Price took a major leadership role in efforts to reduce air pollution in Eugene. In February, 1962 Price, Mayor Edwin Cone and the City of Eugene Wood Utilization and Air Pollution Control Officer, Werner Adkison, had met with lumber industry representatives. On April 9 Price recommended to the Board that EWEB cooperate with the city in solving pollution problems, work with industry through fuel dealers in purchasing wood waste, establish quotas for dealers, negotiate with the Central Heating Company for purchase of its facilities and complete an analysis of the entire program at the earliest possible date. Later that same month the Board approved the steam system expansion and the purchase of the Central Heating Company at a price of $235,000 for all of its facilities. This signaled the Board moving substantially into the area of steam heating service.

". . . WORST STORM EWEB HAS EXPERIENCED ..."

During the morning of Friday, October 12, 1962 the Weather Service was predicting high winds. EWEB operating departments took note of the prediction but as is the case in such matters, other than awareness, little could be done in the way of specific action.

Also on October 12, an all-employee meeting had been scheduled for mid-afternoon. During the course of the meeting, at about 3:50 p.m., the sound of whipping winds could be heard. By 4:30, when the meeting was scheduled to end, the utility operation began to realize what it was confronting. Outage reports poured in and the electrical operations headquarters became a command post.

It was fortuitous that the employee meeting was scheduled and took place when it did. Many EWEB crew members were at the meeting who otherwise might have been in danger areas. If employees had been released early, as sometimes was the case when inclement weather threatened, the results might have been serious.

The writer, possessing the luxury of some flexibility, unable to reach his family by telephone, journeyed to his home—as quickly as possible given the conditions. The normal route to southeast Eugene was by way of Patterson and Hilyard. This being blocked, it was necessary to find open streets in east Eugene and then reach Hilyard by way of 24th Avenue. The sight of the chapel at Sunset Home, which was a heap of rubble, was disheartening. However, the writer's residence, nestled in the lee of Spencers Butte, was undamaged.

Other EWEBers were not so fortunate. Many crew members addressed their awesome task not knowing the status of their families. And EWEB addressed the task. It was a magnificent performance—accomplished without a single lost-time accident. Many EWEB personnel worked through that first night and all the next day before relief.

A legend that grew out of the Columbus Day storm was that of Cliff Rogers and his crew consisting of Gordon Goepford and Bert Pitt. As it was said in those times, the crew was "forgotten" as it disappeared into the up-river service area. Goepford recalls today the assistance rendered the crew by Hydro System Supervisor Darwin Spicer and his crew including Marion "Buck" Carter and Emmett "Fuzz" Lovell. Spicer's wife, Elsie, provided coffee and sandwiches.

Two University of Oregon students reported to EWEB headquarters the first night and volunteered their services. After working all night, the two refused remuneration. Personnel of local electrical contractors joined in the effort.

Five people died in Eugene as a result of the storm—two directly from storm violence and three from other causes. Henry Beistel, quoted in the Sunday *Register-Guard* on October 14, said, "It was the worst storm EWEB has experienced in its 51 years."

DID EWEB ERR?

Customers suffered electric-service outages ranging from a few hours to several days. If an error was committed by EWEB during this experience, this was in the manner it related to its customers. Perhaps in its zeal to serve its customers well, the office staff chose to keep many telephone lines open in order to answer customers' questions about when they could expect service. This created problems. Firstly, it was often impossible to accurately predict when an area or individual customer would be back in service and incorrect information was often given. Secondly, this concern for customers placed a heavy burden on operating personnel who were besieged with questions from the office staff regarding service restoration.

The writer is convinced that the practice followed by other utilities—that of simply announcing to the service area that service would be restored as soon as possible without attempting to predict restoration time for individual customers—was more practical and perhaps of greater ultimate service to customers.

RATES HELD AT LOW LEVEL

During Price's administration the Board was able to hold rates for both water and electricity at comparatively low levels. On April 14, 1969 the Board approved a water rate increase designed to generate an additional $400,000 a year to help meet the cost of a seven-year water system expansion program. On November 23, 1970 the Board's independent engineer, Ford, Bacon & Davis, was directed to study electric rates. On May 10, 1971 an electric rate increase averaging between 15 percent and 16 percent was imposed. There was considerable discussion and debate involved at the Board meeting on May 10. Representatives of the Eugene Homebuilders Association contended that the rate increase would result in greater use of fossil fuels for home heating, thus creating additional pollution problems. The Eugene Future Power Committee, represented principally by John Reynolds who would later become a member of the Board, presented its own rate proposal which was designed to promote conservation and was predicated on load forecasts which differed from those presented by the independent engineer. The recommendation of the independent engineer and the staff prevailed and for the first time in 60 years, a significant electric rate increase was approved. In both the case of water and electricity, Eugene's rates remained very favorable as compared to other rates in the region and the nation.

Throughout Price's administration the Board paid careful attention to wages and fringe benefits for EWEB employees. The utility industry was the principal source of data upon which the Board based its decisions. For a number of years the journeyman lineman wage, which was very nearly uniform on the western side of the Cascade range throughout the region, was the primary benchmark upon which wages were based.

Generally, utility employees throughout the region, both salaried and those receiving an hourly wage, benefited from the increase granted to the linemen. For a period beginning in 1967 the Board differentiated between hourly and salaried employees. Hourly employees were granted wage increases based upon the industry pay scale plan which compared to the general negotiated increase in the industry. Salaried employees were limited to an increase based upon the cost of living which usually was less than the industrial wage increase. However, salaried employees in most cases had available to them periodic step increases within their pay plan.

Also in 1967 significant improvements were made in fringe benefits. A sick leave incentive plan provided 25 percent cash payment of unused sick leave at retirement. Upon the recommendation of Assistant Treasurer William Rau, the Board adopted a program of charging the employee one percent of gross salary as his contribution to the group insurance program. This was a revolutionary step and one which has proven out very satisfactorily over the years. Earlier, on May 1, 1964, long-term disability insurance became available as part of the group insurance program. Thirteen days after adoption of long-term disability insurance a 44-year-old female employee suffered a cerebral hemorrhage and was the first to benefit from the new long-term disability program. The statistics provided by the insurance industry indicating that the possibility of early disability is eight times greater than early death have been borne out over many years and over 30 employees have benefited from the long-term disability provision.

September 9, 1963; EWEB employees, wearing white coveralls and yellow hard hats, help host Carmen-Smith dedication ceremony. (EWEB files, H. Gross photo)

Has EWEB's policy of maintaining industry parity in wages and fringe benefits been sound? The oft-stated simple equation has been that if EWEB had not provided parity, it would have quickly become organized and industry wages and benefits would have been granted through negotiation anyway. A non-unionization stance has never been enunciated by the Board or EWEB management, but maintaining wage and benefit parity probably has been a factor in maintaining a non-union environment. The benefits of this condition probably have been more meaningful and important to the general work force at EWEB than to the management. The reason for this is that in a non-union environment, benefits of greater flexibility accrue to the employees.

EWEB Vice President Calvin R. Smith—among speakers at Carmen-Smith dedication. (EWEB files, H. Gross photo)

POSITION REDESIGNATED "GENERAL MANAGER"

On January 10, 1966 at its annual organization meeting Byron Price was reappointed superintendent-secretary but was also designated general manager. The new title still carried the responsibility of secretary, but it gave much better definition to the chief executive's position and was more in keeping with the titles of the chief administrators of other consumer-owned utilities in the region.

In 1965 Byron Price served as Governor Mark O. Hatfield's coordinator in an effort to locate an atomic accelerator in the state of Oregon. Under Price's direction the staff prepared very elaborate proposal material setting forth the benefits of the state of Oregon as they might apply to the siting of a major accelerator facility. The accelerator would eventually be located in the Midwest rather than in Oregon, but the work accomplished was a very useful exercise in giving definition and dimension to economic development potential within the state of Oregon.

BOARD FAVORS DEVELOPMENT

On May 13, 1968 Board member Ehrman Giustina presented a strong case for assigning economic development responsibilities to Byron Price. On May 27 the Board took formal action on the matter and Price was directed by the Board to acquire adequate assistance so that he might be relieved of much of the day-to-day administration of the utility.

Price's interests moved in another direction early in 1969 when he sought a major appointment in the U.S. Department of Interior under the Nixon administration. Price, an ardent and dedicated Republican, probably could have served the presidential administration well. However, an offer of employment was not forthcoming and the general manager continued his quest to put the Eugene utility directly into the nuclear generating business.

Early in 1966 the Board and General Manager Price sought voter approval for a charter amendment which would permit the issuance of revenue bonds without a vote of the people. It was argued that approval of a measure would permit planning of financing "to fit construction scheduling." Other advantages would be shortening of the time required to obtain financing from seven months to about two months and reduce costs by eliminating elections each time bonds are sold. The concept generally was not a popular one in Eugene and when the election was held on March 15, 1966, there were 2,386 votes against and 1,501 votes for the measure. The *Register-Guard* was opposed to the charter amendment.

Probably the concept of revenue bond issuance without voter approval warrants the consideration of a future Board. However, such a provision should prescribe specific limitations such as debt-equity ratio.

NUCLEAR INVOLVEMENT BEGAN EARLY

The Eugene utility's nuclear involvement did not begin with events of the 1960's. In fact, it began long before the concept of nuclear power was fully developed. In 1933 Dr. Harold Urey, a chemist from Columbia University, visited his friend, Professor O. F. Stafford, head of the University of Oregon's chemistry department. With the help of Stafford and his students and with the cooperation of Superintendent McArthur, Urey was able to carry on his experimentation with the isotopes of hydrogen. Urey was to receive the Nobel Prize for Chemistry for the discovery of duterium oxide (D_2O) in 1934.

Duterium, or heavy water, is an isotope of hydrogen with an atomic weight or mass of 2 instead of 1. It is 25 percent thicker than regular water and 10 percent heavier. Until his arrival in Eugene, Urey had been able to produce only a few grams of heavy water at a time. He needed to find a way to produce a larger quantity of the isotope to help him in his research in the field of atomic energy. Urey told his friend Stafford about the problems he was having. Stafford worked with Joe McArthur to see if an answer could be found.

Several direct current generators lay idle due to the recent discontinuation of electric streetcars in Eugene. Retired Electric Operations Superintendent Vernon J. Poindexter recalls maintaining one of the generators used in Stafford's work. To separate out the hydrogen isotopes the students had to "electrolyze" or "decompose" the elements of water. The utility permitted them the use of a 10,000-gallon water tank to conduct their work. The students had to search for old automobile batteries and use the lead to make electrodes for the electrolysis cells.

Among the participants in the experiment was a young chemistry major named Thomas R. Miles, who is today a consulting engineer in Portland. Miles later wrote of the experience, "Over a period of about nine months as I recall, the 10,000 gallons of water was electrolyzed down to 50 gallons by pumping through the cells, producing large quantities of oxygen and hydrogen gas and a very small capacity of D_2O . . . Eventually the 50-gallon drum of concentrate containing approximately two gallons of D_2O was proudly displayed in the basement of McClure Hall, the then chemistry department building, and then finally shipped off to the National Bureau of Standards to be further concentrated for use by Urey and others."

Scientists viewed this first successful isolation of an isotope of hydrogen to be more important than the discovery of a new element. One writer of the day compared the discovery of D_2O with Columbus' discovery of America.

But at the time it appeared that there was no immediate use for the discovery. The concept of nuclear power was still somewhere in the future and as far as the Eugene Water Board was concerned, there was no connection between heavy water and electric power. Today heavy water is employed in many parts of the world as the moderator in nuclear power plants and thus, at an early time in a most significant way, the Eugene Water Board was involved.

NEED FOR NEW PLANTS STATED

As noted earlier, in 1960 and again in 1961 the Board turned down invitations by the Atomic Energy Commission to participate in power reactor demonstration projects. However, on February 11, 1963 the Board approved executing contracts with Bonneville Power Administration and the Washington Public Power Supply System to obtain power from the nuclear power reactor at Hanford. Also to be considered was an offer by Washington Water Power Company to take its share of power from the Hanford generating plant through 1969. On May 9, 1966 the Board heard a report that one of three nuclear plants which would be needed by 1977 would be built in the upper Willamette Valley. In November, 1966 Price told the Board that nuclear power would be the most likely resource to supplement hydro power.

On July 24, 1967 Price reviewed a report prepared by Battelle-Northwest that two large thermal plants were in the active planning stage—the Centralia coal-fired plant and a one million kilowatt nuclear plant near Rainier to be sponsored primarily by Portland General Electric Company. Price proposed that EWEB study the possibility of joint ownership in the PGE plant and also the possible sponsorship of a large plant "in this part of Oregon." He added that the time had arrived for

Governor Mark O. Hatfield—featured speaker at Carmen-Smith dedication ceremony. (EWEB files, H. Gross photo)

EWEB to embark on a program to provide additional generation. He recommended that the Board commit itself and state its intention to proceed with all facets of planning toward additional sources of power, including thermal nuclear generation, adding that "we have a responsibility to the community to make provision for future power needs."

At the October 9, 1967 Board session, President Calvin Smith reported on meetings held in Boston, New York City and Washington, D.C. with engineering firms and bankers regarding nuclear construction. In addition to the full Board, Byron Price, William Rau and Chief Engineer Charles Strong attended the meetings. Smith said that a 1,000 megawatt plant will be needed and that EWEB should begin its planning. On October 13 the Board approved scheduling a seminar to be presented by NUS Corporation, a nuclear utility consulting firm in Washington, D.C. The seminar was scheduled for November 30 and December 1.

In January the Board agreed to participate in the "This Atomic World" program in cooperation with the University of Oregon, the Atomic Energy Commission and Oak Ridge Associated Universities. Portland General Electric Company and Pacific Power & Light Company were eventually to be involved in this activity. "This Atomic World" was a nuclear power promotion program designed for high school students.

On January 22, 1968 Price told the Board that action was needed to move ahead on participation in up to 200 megawatts of the Portland General Electric Company Trojan plant on the Columbia River—to be followed by an EWEB plant of some 1,000 megawatts in the Eugene area. The Board unanimously approved these concepts with a comment by Board member Kenneth Kohnen that nothing was binding at that point in time. However, negotiations with Portland General Electric continued and on March 12, 1970 contractual arrangements for EWEB's 30 percent ownership of Trojan were approved by the Board.

On March 25, 1968 the Board approved membership in the Atomic Industrial Forum, the principal national nuclear industry political and lobbying organization.

In July 1968, August 20, 1968 was set as the date for discussion of a charter amendment for bond financing in the amount of $225 million for thermal nuclear plant construction. In September, 1968 the Board heard a presentation by Byron K. Taylor on warm water discharge from nuclear plants and its use for irrigation. Following discussion, the Board approved a preliminary study by Vitro Hanford Engineering Service on the matter of warm water irrigation in connection with an upper Willamette Valley nuclear plant.

During the early fall of 1968 the Board conducted a mail-out survey of its customers designed to elicit information on attitudes about nuclear power. The results of this survey indicated that over 82 percent of respondents favored nuclear power with only two percent opposed. Those favoring nuclear power were in evidence at the general election on November 5, 1968 when 21,676 voted in favor of and 5,799 voted against the concept of nuclear power.

Unaware of events which were to take place, the Board late in 1968 moved rapidly ahead with plans to build a nuclear plant. Price outlined a program including staffing and training, site evaluation and selection, consideration of types of reactors and preparation of bid specifications. The Board approved participating in maintaining a nuclear chair at

Carmen-Smith dedication ceremony; left to right; Byron Price, Rev. Wesley Nicholson, Governor Mark O. Hatfield. (EWEB files, H. Gross photo)

Oregon State University at a cost of $6,666 with equal amounts being contributed by Portland General Electric Company and Pacific Power & Light Company.

Other business of the Board was carried forth, and on November 25 air pollution control officer Werner Adkison said that EWEB's part in burning wood waste had contributed substantially to the improvement of air quality in the downtown area of Eugene.

OPPOSITION GROWS

In May, 1969 opposition to the Eugene plant began to crystalize. Members of what was to become the Eugene Future Power Committee began to appear at Board meetings on a regular basis. Notable among the opposition were Joe Holaday, Chris Attneave and Jane Novick.

Notwithstanding the evidence of local opposition to construction, the Board engaged Stone & Webster as its architect-engineer on June 23. At the October 27 meeting of the Board a large group of Eugene businessmen appeared and presented comments favoring the Eugene plant. At a later meeting a letter was read from Joe Holaday, as chairman of the Eugene Future Power Committee, in which Holaday requested information on the "so-called warm water project" and answers to other questions about assistance to institutions and individuals in the field of education, presumably referring to EWEB's participation in the Oregon State University nuclear chair. In December, 1969 the Board directed that the following manufacturers be asked to submit proposals: Babcock & Wilcox, Combustion Engineering, General Electric Company, Gulf Atomic General and Westinghouse Electric Corporation. Types of reactors to be considered were both boiling- and pressurized-water reactors and the high temperature gas-cooled reactor which was being planned by Gulf Atomic General. At an earlier meeting, Florence, High Prairie—located near Oakridge, Oregon and Poodle—located at Poodle Creek west of Eugene, were recommended as possible sites by Kenneth Sieving of Stone & Webster Engineering Company.

During the winter of 1969-70 the Eugene Future Power Committee launched an initiative campaign which would impose a four-year moratorium on the construction of a Eugene plant. When sufficient signatures had been obtained the City Council on February 24, 1970 approved placing the measure on the ballot to be voted on at the primary election on May 26.

The ensuing campaign rivaled the 1932 Zorn-McPherson initiative as a dramatic if not traumatic experience for Eugene. The major difference was that while Zorn-McPherson brought the community together as a solid front against a threat from outside, the nuclear issue tore it apart. The *Eugene Register-Guard* allied itself with the Eugene Future Power Committee and was instrumental in the success of the moratorium election. At the election on May 26, 1970 the moratorium was approved—with 11,750 voting yes and 10,892 voting no.

During the height of the nuclear fight and as opposition to the plant mounted, former Board member and University of Oregon Law School Dean Orlando Hollis and University of Oregon physics faculty member Russell Donnelly announced formation of SCAP, which stood for Scientists and Citizens for Atomic Power. Donnelly was president of the group, Hollis vice president and former mayor Edwin E. Cone, treasurer. Another pro-nuclear organization was CODE which was formed earlier and was the acronym for Citizens for the Orderly Development of Electricity.

The opponents of the plant generally were intelligent, articulate and well-grounded in their homework. Similarly, proponents of the plant, including the members of the Eugene Water & Electric Board, were intelligent and articulate in their presentations to many organizations.

In retrospect, after 15 years, it is neither possible nor appropriate to assign culpability to any individual or organization involved in the 1970 campaign. The members of the Eugene Water & Electric Board were dedicated men sincerely interested in the best interests of the community and economic development. Perhaps they were not mindful enough of the activist movement of the 1960s and growing national concerns about nuclear power. At the same time, opponents of the plant sincerely believed in their position and certainly it must be added that power-needs forecasts developed by the Eugene Future Power Committee proved to be more accurate than those which were being set forth by the utility industry.

Whatever the case, Eugene was a microcosm of what was taking place throughout the nation. Citizens everywhere were questioning authority and demanding to be heard. They were no longer content to leave important decision making to political bodies.

The Eugene Future Power Committee also addressed EWEB's involvement in Portland General Electric's Trojan nuclear plant on the Columbia River. On March 12, 1970 members of the committee told the Board that inasmuch as Trojan power would not be available for 14 years, until June 30, 1984, it could not possibly meet the power needs that EWEB claimed it would have in the mid-1970s.

It was the intention of General Manager Price and the Board to keep the option open for nuclear plant construction at the Big Creek site located 12 miles north of Florence. The Board was prompted to abandon the upper-Willamette Valley during the moratorium campaign and the coastal site seemed the most acceptable choice at the time. A letter of intent was filed with the new state Nuclear Thermal Energy Council (N-TEC) predicated upon the concept that the plant would be built following the four-year moratorium.

At the urging of the Eugene Future Power Committee, the League of

Women Voters and others, the Board agreed in November, 1971 to print an energy conservation brochure to be mailed to all customers. At the same meeting Price explained the liquid metal fast breeder reactor program and although no action was taken, the Eugene Future Power Committee at a subsequent meeting requested that for whatever funds might be allocated to the fast breeder reactor program, the Board contribute an equal amount to a program for slowing down increasing demand for energy and promoting research and development for alternative resources.

The Board did not heed the Future Power Committee's request, and at its meeting on December 9, 1971 it approved an allocation of $35,800 per year for a period of 10 years as a contribution to the liquid metal fast breeder reactor program. This was to become known as the Clinch River Project.

OBJECTION TO CHAMBER MEMBERSHIP

In March, 1972 members of the Future Power Committee raised the issue of General Manager Price attempting to bring new industry into the Eugene area. They were told that although he had not been active in this area, Board action directing Price to pursue such activities remained in force. At the same meeting members of the Future Power Committee also objected to EWEB's participation in the Eugene Chamber of Commerce. To this the Board said that it would consider the matter at its next budget meeting. When the budget was considered in November 1972, financial participation in the Chamber of Commerce was reduced from $10,000 to $5,000. However, in December the Board expanded its participation in chamber activities and approved a total of $15,580 which included $12,850 for contract services with the Chamber and $3,000 of EWEB staff time to be used in conjunction with Chamber activities. This action was followed by discussion on the Board's involvement in socioeconomic development.

Interest had been developing in alternative resources. Dr. Wendell Hewson, chairman of the Department of Atmospheric Sciences at Oregon State University, asked the Board for a grant of $25,000 for wind research. Newly elected Board member John Reynolds was immediately prepared to move for approval of the proposal. But the remainder of the Board was not yet ready to take action. Also Dr. Lloyd Staples, geologist and consultant on geothermal energy, had told the Board that there were promising resources within the state of Oregon.

Discussion about nuclear energy continued, and in January, 1973 Dr. Ray Wolfe, University of Oregon chemistry professor, raised the issue of the emergency core cooling system at the Trojan facility, questioning the safety of the proposed design.

Following the May 26, 1970 moratorium election the Eugene Future Power Committee appeared to have assumed the role of watchdog over the activities of the Board. This would inevitably lead to conflict. On February 26, 1973 Board Member Richard Olson admonished Joe Holaday to "take a position" rather than besieging the Board with unceasing questions. This condition was most difficult for Board members who were attempting to represent their constituencies in the best way they could. Ehrman Giustina had already resigned and Board member Richard Olson was to resign in May, 1973.

In March, 1973 General Manager Byron Price told staff members that he intended to resign from his position with EWEB. On July 10, 1973, at an emergency Board meeting, Price's retirement was announced effective August 31. Keith Parks, who had been appointed assistant general manager earlier in the year, was appointed acting general manager. Later in July the Board discussed with its independent engineer, Ford, Bacon & Davis, criteria for the selection of a new general manager. Also on the Board's agenda was consideration of the recently passed open meeting law, as well as indemnification for Board members and top level employees in activities other than cases of malfeasance in office or willful or wanton neglect of duty.

Also during the summer of 1973, because of the critical water condition in the Columbia River system, power shortages appeared to be imminent and the Board was quick to respond with local conservation efforts as well as cooperating with statewide conservation programs.

Late in July, Herman Kehrli, League of Oregon Cities, appeared at a Board meeting and asked to be consulted in the selection of a new general manager. The Board heard Kehrli but decided to utilize the services of Ford, Bacon & Davis in its quest for a new administrator.

In September the Board learned that the cost of the Trojan plant had increased from $298 million to $345 million with commensurate increases involved in EWEB's 30 percent share.

Although heavy rainfalls throughout the region in the fall of 1973 solved the problem of an immediate energy shortage, conservation was soon to become a primary activity among utilities in the Pacific Northwest. EWEB efforts along these lines included additional customer information and the work of an employee committee which was to be commended by the Board.

PARKS APPOINTED

After a nationwide search for a new general manager, the Board turned back to its incumbent staff and appointed Keith Parks as general manager on January 9, 1974. In a subsequent meeting on January 14 Byron Price commented, "Parks was one of the finest administrators I have ever known and the Board putting him in this role makes me very proud."

In an interview during 1976 Byron Price reflected upon his role as administrator of the Eugene Water & Electric Board. He spoke of the professionalization of the management of the Eugene utility operation, giving particular credit to Henry Beistel in the fields of general administration and personnel management; to John E. Brown, who as treasurer helped divorce EWEB from the apron strings of the City Council; and to Byron Taylor, who successfully coordinated the work of the engineering and operations departments.

He was very proud of the downtown underground electric network, giving particular credit to Jim Howell, Mel Damewood, Don Robinson, Dean Axtell and all of the engineering department; to Vernon J. Poindexter and Glenn Haney of electric operations; and to George Partridge of the construction department. Eugene's underground network system attracted attention nationally and a process was developed whereby a pre-cast underground vault could be set in place with excavation, installation and covering taking place in a single day.

In 1972 Price had assigned Edith Warner the task of establishing compliance with civil rights laws and the development of an affirmative action program. Well before that time EWEB had made strides in the employment of minorities.

Price rejected the traditional hard-line utility adage that "not a single kilowatt-hour can be extracted from a blade of grass," and undertook an extensive facility beautification program. Substations and reservoirs alike, under the guidance of Victor Kuhlman, were beautifully landscaped.

Price was particularly proud of the water system and the method of superchlorination followed by dechlorination designed to protect customers from any kind of waterborne disease. He was, however, disappointed that fluoridation had remained a very emotional issue in Eugene and could not be implemented.

EWEB had played an important role in alleviating a serious air pollution problem in Eugene. Price attributed much of the success of woodwaste burning efforts to Owen Brown and Clark Lynch, who devised means of burning a low-quality-wood waste product.

Price shared the frustration of Henry Beistel in the inability of the utility to achieve the conversion of existing residential overhead electric systems to underground. Even though Beistel was instrumental in having municipal utilities included in a statute permitting bancrofting* of such conversions, there were no takers, and conversions to un-

*"Bancrofting" refers to a method of financing public improvements in Oregon. Fred A. Bancroft was a member of the 1893 Oregon Legislature. Legislation which he introduced provided that cities of 2,500 or more population could issue general obligation bonds for the improvement of any street or part of a street or for the laying of sewers—with debt retired by assessment against the benefiting properties. The legislation, which has been amended many times, has served as a model for other states.

derground when cost was involved simply did not appeal to customers. However, under Price's administration the Board adopted a policy of undergrounding all new facilities and many of the newer residential areas of the city are served by an underground system.

Price had been frustrated by the emergence of professional planners and super-agencies at the federal and state levels which created a "foreign dictate." He said that some planners were theoretical only and caused great problems. He credited Keith Parks during his then short tenure as general manager as having been able to sort out many of the complexities of relationships with other governmental agencies, and said that he had "been outstanding in relating to various planning and other agencies."

He credited Miller Evans, Herb Hunt, Hal Worcester, John Fuller and Barbara Lamb with efficient management of power resources and power purchase contracts and arrangements, and said that the groundwork was being set for eventual western states power coordination and eventually national coordination.

MORATORIUM "DEPLORABLE"

Price said in 1976 that the nuclear plant moratorium took Eugene out of nuclear as well as other large-scale thermal plants. He considered the moratorium as deplorable and that it eroded the credibility of EWEB in its role as a leader. He considered local opportunities for generation such as the Weyerhaeuser-EWEB Energy Center as commendable, but not the answer to future energy needs.

Byron Price made a significant and lasting contribution to the city of Eugene and its utility system, but in the later years of his administration he was caught in the clash of conflicting ideologies. He had gone to work for a Board which had been solidly representative of the business community for 46 years. By 1970 a new condition was solidly in place—creating an environment in which Price found it very difficult to work. Board meetings no longer consisted of simple, straightforward, orderly handling of utility affairs. Rather, meetings were punctuated by lengthy dissertations which seemed to have little connection with traditional utility concerns.

The process which took place during those times cannot be faulted. It was in keeping with Board philosophy which encourages public input. And certainly the stakes were changing. The cost of large, thermal generating stations was accelerating at an incredible rate. Concerns about nuclear power were growing throughout the nation—not only in Eugene. The nuclear moratorium election in 1970 was a proper exercise in the American political process and it brought about change in the general direction of the Eugene Water & Electric Board.

The two major disappointments in Price's tenure as general manager were loss of the nuclear plant and his failure to obtain a major appointment in the Nixon administration. Notwithstanding these disappointments, Price was largely responsible for Eugene's utility system achieving national recognition and acclaim. He can be proud of his efforts.

CHAPTER SEVEN
Keith Parks—Time of Change

KEITH PARKS ASSUMED ADMINISTRATIVE CONTROL of the Eugene Water & Electric Board following a year of confusion, if not frustration, in the utility industry. During the spring of 1973 low-water conditions in the Columbia River seemed to portend what could be severe shortages of electricity in the fall of that year. However, heavy rains in November raised the reservoir levels to a point where the danger of imminent shortage was eliminated and the Bonneville Power Administration, by the end of 1973, was issuing news releases about energy surpluses in the region. Northwest utility customers were receiving contradictory messages about possible energy shortages and the claim by BPA and the utilities that several additional large electric generating plants would be needed raised doubts. At the local level, reporters representing some of the electronic media, despite patient explanation by EWEB spokespersons, were unable to distinguish between low-water on the Columbia River and Eugene's domestic water supply. They told listeners that water conservation was needed in Eugene when such was not the case.

Keith Parks, like his predecessor, Byron Price, did not possess the formal credentials of earlier administrators. He was without a college degree but like his predecessor, he was indefatigable and of high intellect—qualities which would compensate well for any deficiency in credentials.

DIFFERENT KIND OF BOARD

The Board which hired Keith Parks departed considerably in composition from that which employed Byron Price 17 years earlier.

John A. Tiffany, president of the Board, was president and general manager of Tiffany-Davis Drug Company. He had served as a member since his appointment in 1962. Richard F. Freeman, vice president, had been elected to the Board in 1970. He was a geology-research consultant at the University of Oregon and owner of a local motel. Taylor Ramsey had been appointed in October 1972. He was president of United Pipe & Supply Company, Inc. and Ramsey-Waite, Inc., and their respective affiliates in Portland and Medford. John S. Reynolds had been elected for a four-year term which would expire in 1976. Reynolds was an associate professor of architecture at the University of Oregon. Calvin L. Schmidt had been appointed in June 1973 and was co-owner of the industrial building firm of Schmidt & Smith, Inc.

While Tiffany, Ramsey and Schmidt represented the more traditional business component of the community, departure from the historical mode was represented by Freeman and Reynolds. Freeman, by January 1974, had already demonstrated his independence of thought and the fact that he was a researcher and observer first and businessman second. John S. Reynolds, a member of the Eugene Future Power Committee, had demonstrated his intelligence and astuteness in grasping utility problems. Tiffany, a veteran of nearly 12 years' utility governing board experience, provided the wisdom of that experience and efficiently managed the conduct of Board affairs through his role as president. Taylor Ramsey and Calvin Schmidt were successful entrepreneurs who added their business acumen to that of President Tiffany in fostering a businesslike approach to utility affairs.

Age 48 at the time of his appointment as General Manager, Keith Parks possessed a rich background in local government. His employment with Lane County government dated back to brief service before his entering the U.S. Army Air Corps during World War II. He returned to county government after the war and rose through the ranks to become the first director of general administration under the Home Rule Charter on July 1, 1963. He joined EWEB August 19, 1968 as assistant to General Manager Byron Price. Early in 1973 he became assistant general manager and upon the retirement of Byron Price on August 31, 1973, Parks became acting general manager.

The top management staff which Parks fell heir to included Herbert H. Hunt, who was serving as director of operations & engineering. Hunt, age 49, had been employed by EWEB on September 3, 1968. He had graduated from the University of California at Berkeley in electrical engineering and, prior to coming to Eugene, had been employed by the Sacramento Municipal Utility District for 18 years. John E. Brown, age 50, was treasurer and director, accounting & finance division. He joined EWEB on October 30, 1967, following 17 years with Commonwealth Services, Inc. in Jackson, Michigan. He held a bachelor of business administration degree from the University of Minnesota. William F. Rau, 47, director - general services division, had been with the organization since February 14, 1966. He graduated from the University of Oregon School of Business in 1949 and, prior to coming to EWEB, he had broad management experience in the insurance industry. Norman F. Stone, 48, director - administrative services division, joined EWEB as personnel officer on May 1, 1956. Stone held degrees from Willamette and Stanford Universities and, prior to coming to EWEB, spent four years in Oregon state government.

Richard D. Gall, 47, was electric operations superintendent. He had joined EWEB in May 1950 and through rigorous self-education had worked up through the ranks of electric operations. Kimber G. Johnson, 34, was water operations superintendent, a position he had held since December, 1971. Johnson joined EWEB following graduation in civil engineering from Brigham Young University in 1965. He would subsequently receive a master of business administration degree from the University of Oregon. Owen Brown, 48, was steam system superintendent. A mechanical engineering graduate of Washington State University in 1950, Brown had joined EWEB in September 1965. Clarence Hill, 56, was hydro system supervisor. He joined EWEB in February, 1965 following long service with the federal government.

Temporarily out of the management group was Kenneth W. Rinard, 47, who had served many years as water operations superintendent. Rinard had been assigned as an information service system analyst to help develop a management information system for EWEB. In January, 1975 he would return to the management team as assistant director - engineering and operations and would prove to be an important influence in Parks' administration.

The position of chief engineer was not officially occupied at the time of Parks' appointment. These duties were carried by Herbert H. Hunt until the appointment of Owen Brown to the position which would take place in January 1975.

During Byron Price's tenure as general manager-secretary, annual electric system revenues increased from $3.5 million to over $15 million, an increase of 328 percent. Water revenues increased from about $621,000 to nearly $2 million, an increase of 222 percent. Average electric system customers increased by over 72 percent, averaging 48,541 during 1973. Average water customers increased to 24,309, amounting to a 118 percent increase. The larger percentage figure for water customers was brought about by the acquisition of additional customers through annexation of territory by the city of Eugene. While this increase in utility services was taking place, the work force increased from 265 to about 380 at the end of 1973, a gain of only 43 percent.

DIFFICULT CHALLENGE

The new general manager-secretary faced the task of conducting the affairs of the utility, but perhaps his most important challenge was that of healing the wounds which remained from the events of the latter part of the previous decade and particularly the nuclear fight of 1970. On a larger scale, the entire electric energy establishment in the Pacific Northwest had lost credibility. The major issue was not the struggle between private and public power. Rather, it was regaining a working relationship with a citizenry that had, for considerable reason, lost a great deal of confidence in the energy bureaucracy.

As part of the rehabilitation process, Parks realized that he must have broad-based community support. He faced the problem of moving the operation of the utility forward, but he also sought and achieved the support of the Eugene Future Power Committee which had developed such an adversarial relationship with the former general manager. Parks also found a valuable ally in the *Eugene Register-Guard*.

Early in 1974 Commissioner John Reynolds urged a broadening of the Board meeting process, including "special process" meetings every three months. The other Board members felt that the present Board meeting structure was adequate and permitted a proper level of public input and Reynolds' proposal was never adopted.

Members of the Future Power Committee, including Leslie Cox, Marian Frank, Chris Attneave and Fred Keppel, continued to attend Board meetings and provide input about such matters as the visitors information center at the Trojan nuclear plant. Dr. Ray Wolfe, University of Oregon chemistry professor, invariably prefaced his remarks with a statement that he was representing himself only and was not associated with an organization. His contributions to Board discussions were often thoughtful and substantial.

GOOD PERSONNEL PRACTICES CONTINUE

Parks, as did his predecessor, paid close attention to the wage settlements in the utility industry. On April 29, 1974 the Board approved an eight percent interim wage increase to be effective on May 1, with the provision that later contract settlements would be taken into consideration. In July, when the industry wage pattern had become established, the Board acted to revise its earlier eight percent wage adjustment to 11.03 percent.

Earlier in the spring of 1974, Board members Taylor Ramsey and Cal Schmidt were appointed as a special committee to study EWEB construction practices. In June of that year they reported that, in general, the work force was efficient and made good use of equipment. However, they recommended the operation utilize computerized manpower and cost information to achieve better cost control.

In July the Board approved expansion of the Hayden Bridge Filtration Plant as an alternative to developing a new water source elsewhere on the McKenzie River or on the Willamette River. At the same meeting the staff was directed to secure options on sites on the McKenzie near

Armitage Bridge or on the Willamette in Eugene as possible future sources of water. It is interesting that after many decades in which only the McKenzie River was considered a viable water source, the municipal system found that the Willamette River could again be looked upon as a satisfactory water supply.

Keith Parks, EWEB general manager-secretary 1974-1984; guided organization through period of greatest change. (EWEB files)

During the summer of 1974 Commissioner John Reynolds frequently suggested that water consumers place a brick in their toilet tanks as a water conservation measure. There was division of thinking on the Board over this matter and Commissioner Taylor Ramsey often took issue with Reynolds—pointing out that there was not a shortage of water in Eugene.

Walter Youngquist, the Board's geothermal consultant, reported to the Board in August 1974 that there were promising geothermal sites in the Cascade Range. At the same meeting Harold Worcester reported on the progress of the Raft River geothermal demonstration project in Idaho.

NUCLEAR OPTION ABANDONED

While the Board was looking in several directions for alternative energy sources, it took additional action toward backing away from nuclear energy. In November 1974 the Board abandoned options for property on the Oregon coast which would have been needed for the Big Creek project. The staff recommendation stated that the Eugene utility had "no viable arrangements for partners in either the public or private sectors, and it is not in a position now nor will it be in the near future to sponsor such a monumental project." It was also recommended that the letter of intent filed with the Nuclear and Thermal Energy Council for the Big Creek project be withdrawn. This action was approved and thus those in Eugene who favored nuclear generation now had to resign themselves to indefinite postponement of such a project.

As early as September 1974 Commissioner Reynolds encouraged inviting two anti-nuclear spokesmen to Eugene to speak to the Board and others in Eugene. His motion to ask Carl Hocevar and J. M. Hibbard offered at the October 19 meeting died for lack of a second. Reynolds again raised the issue at a November meeting and, after a great deal of debate, particularly between Commissioners Reynolds and Ramsey, Tiffany directed the staff to contact Hocevar and Hibbard regarding their availability and the cost involved. Finally on December 9, following receipt of letters from Hocevar and Hibbard in which they expresed difficulties in including a Eugene visit on their schedules, the matter was formally dropped by the Board with a vote of 3-1. At the same meeting General Manager Parks suggested that the Eugene municipal system should terminate its payments to the Clinch River liquid metal fast breeder reactor project. This action was later taken.

Periodically, during 1974, Commissioner Reynolds, Professor Wolfe and others raised the issue of inverted, or conservation, electric rates. By this was meant a rate form in which the first block or blocks of energy would cost the customer less per kilowatt-hour than the blocks of energy occurring later in the consumption pattern. The term "inverted rate" meant the opposite of the declining-block rate which historically had been favored by many electric utilities. The declining-block rate, according to its opponents, tended to favor those who consumed large amounts of electricity and also it was thought that such rates encouraged the waste of energy. The flat rate for electricity pricing which came into being during the 1970s at EWEB tended to be a satisfactory compromise. It consists basically of a non-commodity customer charge and a uniform kilowatt-hour charge applicable regardless of the amount of energy consumed. This rate form tends to penalize the large user insofar as large consumption carries a concomitant higher charge for electricity used. The Board at its meeting on December 11, 1974 approved a 12.01 percent electric rate increase. Commissioner Reynolds, still favoring a conservation rate, cast the single negative vote. At the same meeting the swimming pool special electric rate was eliminated and it was announced that the church rate would be taken under consideration by the Board.

EWEB, departing from its historical mode, developed a keen interest in the early 1970s in the matter of how electricity is used on the customer's side of the meter. Survey work had been accomplished on conservation attitudes and in January 1975 a questionnaire was approved, designed to learn more about senior citizens, their economic condition and their attitudes about the use of electricity.

"SILENT SERVICE" NO LONGER ENOUGH

What this all meant was that "silent service" was no longer enough. Two factors were basically responsible for bringing about the change in the role of EWEB and electric utilities everywhere. Firstly was increasing consumer awareness with regard to all major institutions in society. Secondly was the increasing price of the commodity involved. The old "one cent club," meaning the price of a kilowatt-hour at less than one cent, at one time the pride and boast of public power in the Pacific Northwest, was a thing of the past. Utility bills were becoming an increasingly significant part of the household budget and it was necessary for utilities to respond accordingly.

On February 10, 1975 the Board acted to eliminate the application of the residential electric rate to churches within its service area. In December, 1955 EWEB commissioners had approved a rate classification change for all churches and church-supported buildings from the commercial to the residential schedule. The minutes of that meeting do not reflect any particular comment or discussion. Investigation by staff in the early 1970s revealed that Eugene was unique in extending a special favorable rate to churches and church facilities. Perhaps the Board that existed in 1955 felt that the special concession which was made to churches was in the best interests of the community, but the results of that action proved to be damaging to churches and to the relationship between EWEB and the leadership associated with the various churches.

With the residential rate in place without a demand factor included for most of the time period involved, churches were built and operated without taking into consideration a demand charge. Thus, by 1974, many churches were demonstrating a consumption pattern with a very low load factor. This is to say, a typical church would have a high kilowatt demand, perhaps at most a few times a month, with a very low consumption during the remainder of the month. The Board did all that it could to ease the trauma of the change. Parallel metering with both straight kilowatt-hour consumption and consumption with demand factor included were employed in order to inform church leadership about the impact of the change.

During staff discussion of the proposal to make the change it was pointed out that other "not-for-profit" organizations such as United Appeal, Elks, Moose and Eagles, as well as tax-supported governmental agencies, were on the commercial schedule and that there was no basis for exempting churches from the schedule. In the transition, demand meters were installed at 121 churches beginning in April 1975 and comparative residential and commercial service billings were sent to all churches through April 30, 1976.

COMMITMENT TO CIVIL RIGHTS

In March 1975 Board members and General Manager Parks signed a statement of commitment to civil rights. The affirmative action program had been put in place through the efforts of Edith Warner and the Eugene utility operation was far ahead of most other organizations in pursuing its legal and moral responsibilities in this new chapter of American history. Key staff members attended comprehensive seminars on civil rights matters and in-house training was provided to all levels of management and supervision. While making important strides in the civil rights field under the guidance of Warner, EWEB continued to recognize the needs and aspirations of its work force in general. The advancement of career employees to positions of responsibility was continued, if not enhanced, under the administration of Keith Parks. Employee input continued to be encouraged. Perhaps it was the continuing absence of a union contract with its attendant constraints brought about by labor-management polarization that permitted the easy flow of information through various levels of the organization.

During the spring of 1975 issues being considered by the Legislature in Salem occupied the attention of the Board. Glen Stadler, former state senator and EWEB's lobbyist in Salem did an outstanding job of protecting the interests of the municipal-utility operation and its customers.

An effort was being made to enact legislation which would include EWEB under the authority of a state contract review board. This represented a perennial type of legislation which, in one way or another, confronted the Board at every legislative session. The Legislature was also considering the establishment of a state department of energy which would include the Nuclear and Thermal Siting Council. In May, 1975 Stadler reported to the Board that Senate Bill 483, establishing the Department of Energy, apeared to be in EWEB's best interests.

In May 1975 the Board approved a 12.2 percent wage increase, again on an interim basis until labor contracts among Northwest utilities had been finalized. In September the Board increased the wage adjustment to 13.8 percent effective retroactive to May 1.

YEARS OF WORK TO NAUGHT

For several years the EWEB staff had been working with the Lane County Electric Cooperative with regard to customers living in a "no man's land" who were not being served by the appropriate utility with respect to the boundaries of the city limits. Upon reaching agreement satisfactory to both utilities, the matter was presented to the EWEB Board and gained its approval on June 9, 1975. The agreement required approval of the Oregon Public Utility Commission. In July the PUC, in its order, declared that the two utilities could move ahead with the exchange of customers with the provision that each customer could choose the utility which he wanted to serve him. This PUC order, which constituted an administrative and operational impossibility, was thrown out by the EWEB Board with similar action taken by the governing board of Lane County Electric Cooperative and thus years of hard work by staff members in both organizations were to naught.

BOARD MOVES ON COGENERATION PROJECT

In his annual message on January 13, 1975 President Tiffany had announced that EWEB would study the possibility of joining the Weyerhaeuser Company in the construction of an industrial-energy center at the company's Springfield plant.

On June 9, Herb Hunt, director of engineering and operations, told the Board that preliminary work had been done and that the project could proceed. Hunt said that the estimated cost was $8.5 million and that financing could be accomplished through the sale of thermal generation revenue bonds approved by the voters in 1968.

The concept of using industrial-waste heat for the generation of electricity, referred to as cogeneration, was considered a promising energy resource and EWEB was one of the first utilities in the region to undertake such a project.

Hunt explained that the 51 megawatts of output at the facility would not be needed by EWEB until 1984 but that it would be prudent to proceed at that time because of sharply escalating costs of construction. He added that until the power is needed by EWEB, it would be sold to other utilities.

The Board ratified the Energy Facility Site Certification in August and after successful completion of the construction schedule, the plant was ready to go on line in November, 1976. The projected cost was $1.3 million less than the original estimate of $8.5 million.

UNDERGROUND RATE ELIMINATED

In October, 1975 an important step in the EWEB electric-rate structure was taken when the R-7 rate for underground service was effectively eliminated. This change provided that the $3.40 per month which had been charged for underground service would be credited against a total of $280, with any excess credited back to the customer. This action was based in part on a presentation on underground policies by Chief Distribution Engineer Mel Damewood in September. Damewood set forth the premises that first of all there should be fairness and equity among various classes of customers, that uniform and consistent policies and procedures be employed, that "front-end" charges for the excess cost of undergrounding be paid by the benefiting parties, and that an average excess cost for underground service be calculated and applied. The action in October eliminating the underground rate was a progressive step in EWEB's electric rate structure and customer relations.

Due in large measure to the excellent work accomplished by Glen Stadler in Salem during the 1975 legislature, EWEB escaped mandatory inclusion in a state public contract review board and in November, the Board approved naming itself as its own contract review board.

Board discussions during 1975 had not been totally harmonious and in December, during the absence of Commissioner Reynolds, he was taken to task by Commissioners Schmidt and Ramsey for alleged statements inferring that the EWEB Board had agreed to stay out of nuclear generation. At the first meeting of the Board in 1976, Reynolds, angered by the statements of Schmidt and Ramsey, asked all of his fellow Board members if he had been rebuked. After some rather heated discussion the matter appeared to be resolved but disharmony was to continue at the same meeting because of the Board's refusal to hold hearings on a nuclear safeguards ballot issue. This measure was vigorously discussed at subsequent meetings of the Board. In March ratepayer Camilla Pratt questioned EWEB's membership in the Atomic Industrial Forum, particularly the legality of EWEB contributing to the Forum's role as a lobbyist for nuclear power. In response, General Manager Parks said that EWEB's contribution was for technical publications and not for the Forum's public information program.

By the spring of 1976 the nuclear option was beginning to be foreclosed not only in Eugene but for utilities across the land. The oil embargo in 1973 had been a harsh reminder to the world that fossil fuel resources were not only finite, but vulnerable to political and economic forces as well. Information about dam-caused nitrogen harm to the sal-

EWEB-Weyerhaeuser Energy Center, employs waste heat to generate electricity. (Northwestern Photographics)

mon fishery had entered the public arena and it was unlikely that large hydroelectric structures would be added to the Columbia River system. And there was grave concern about preservation of the *biosphere*, that fragile film of life support described by Arnold Toynbee in his final volume, *Mankind and Mother Earth,* published in 1976. The Board moved to respond to these changing conditions. The Eugene system was to become a leader and pioneer in the areas of conservation and renewable resources.

Following what had become a pattern, in May, 1976, the Board granted a seven percent wage increase on an interim basis and agreed to consider the matter again when industry settlements had been reached and accordingly, finalized a 7.33 percent raise in June. At the same time, dependent dental insurance was added to the employee benefits program as was life insurance in the amount of 2½ times annual salary.

"ARKANSAS HOUSE" ADOPTED IN EUGENE

During 1976 an EWEB team consisting of Jack Smutz, Dave Coon and consultant Ray Wiley visited the state of Arkansas in order to investigate the "Arkansas House" program. This construction concept, which featured all of the latest innovations in insulation, was adopted by the Board, and in August, 1976, the first certificate was granted to J. M. Steinmuller, Jr. for a house which had been built by Modena Homes. The energy-efficient home program was an example of EWEB's advanced efforts in the field of conservation. Other utilities in the Northwest and the Northwest Public Power Association looked to Eugene as additional energy-efficient home programs were instituted.

Early in 1976 EWEB launched its water-system-improvement program. A water system improvement revenue bond election for the sale of $15 million in bonds had been successful the previous November, and the program envisioned would assure an adequate water supply for the city through the end of the century.

Prior to the November, 1975 election, the EWEB staff task force did its work under the close direction of Board Counsel Windsor Calkins. An earlier event influenced Calkins' guidance and in a letter to General Manager Parks on September 5, 1974, Calkins cautioned:

> It is the law of this state since the case of Porter vs. Tiffany that the Eugene Water & Electric Board has no expressed or implied authority to expend public funds for campaign expenditures.
>
> Therefore no expenditures should be made for radio, television or newspaper advertisements, however factual, concerning the coming election. Information concerning the election should not be placed in monthly bill envelopes.

Calkins' reference to Porter vs. Tiffany was in connection with a lawsuit brought by Eugene attorney Charles O. Porter against Board members in office during the 1968 and 1970 nuclear-related elections. In addition to Tiffany, the suit involved Commissioners Ehrman Giustina, Kenneth Kohnen, Richard Olson and Trygve Vik. The suit, filed in 1970, charged that the incumbent Board members had spent money improperly during the two election campaigns. Lane County Circuit Court Judge Edwin Allen declared that the five Commissioners were not personally responsible for the approximately $13,000 involved. However, the Oregon Court of Appeals reversed the lower court and when the Oregon Supreme Court was petitioned to review the finding of the Court of Appeals, it rejected the petition and the Board members were compelled to pay. Calkins did not agree with the outcome but he accepted it as a significant guideline for the conduct of Board affairs.

In spite of the constraints which had been imposed by Porter vs. Tiffany, EWEB staff members were able to acquire public service time in the electronic media as well as speaking engagements with a variety of groups, and thus the appropriate information was given to the public without the expenditure of money. Notable among the task force members were Earle Clark, Dave Coon, Bill Rau, Ken Rinard, Jack Smutz, Eileen Kunkel, Mel Damewood and Glen Stadler. Water Operations Superintendent Kimber Johnson also contributed a great deal through personal appearances made on television.

ANOTHER ATTEMPT AT FLUORIDATION

The issue of fluoridation again was raised in Eugene and was approved by voters on January 3, 1977. As had been the case previously, this caused a great amount of reaction among many citizens and the matter, for practical purposes, was becoming academic in Eugene. Fluoridation would again be voted out and it seemed unlikely in 1977 that fluoridation was in Eugene's foreseeable future.

In his annual message in January 1977, President Tiffany called for retrofit insulation programs, further geothermal exploration, a continuance of the water improvement program, the investigation of solar energy and low-head hydrogeneration and careful scrutiny of the legislative process.

UNEQUAL SHARE OF PREFERENCE POWER

In 1977 there was considerable discussion in Salem about the fact that the state of Washington was receiving a far greater share of BPA preference power than was Oregon. Some 80 percent of residential customers in the state of Washington were served by public utility districts and municipalities, and thus were the ultimate beneficiaries of preference power. In Oregon there existed the converse in which 80 percent of the residential electric customers were served by investor-owned utilities and thus not recipients of preference power. The benefits of preference power are based on the concept that consumer-owned utility organizations have first call on the relatively low-cost power marketed by the Bonneville Power Administration. While investor-owned utilities were able to meet some of their needs through purchases from BPA, they also had to rely upon other power sources which were considerably more expensive than Bonneville power.

The 1977 Legislature, at the urging of Governor Bob Straub, enacted the so-called DRPA bill. This legislation would establish the Domestic Rural Power Authority which would provide a mechanism for investor-owned utilities to sell preference power to their domestic and rural residential customers. Because of later enactment of the Pacific Northwest Energy Act, the DRPA legislation did not go into effect. The federal legislation would provide an exchange mechanism whereby investor-owned utilities could acquire preference power for their residential customers, thus accomplishing the essential objective of Governor Straub's bill.

Power Resources Engineer John Fuller took the lead in developing EWEB'S mandatory curtailment program. EWEB's plan, which eventually would meet the standards established by the Oregon Public Utility Commission, provided for several stages of curtailment which eventually could involve penalties for those customers who did not meet curtailment requirements.

CONSERVATION CENTER ESTABLISHED

In May 1977, John E. Scofield, longtime EWEB electrical engineer, was appointed EWEB energy conservation manager and separate offices for a conservation center were established in the Oregon Building located on Willamette Street. Scofield accomplished an outstanding job organizing EWEB's conservation department. By August 1977 he reported to the Board that an industrial energy advisory program and the energy-efficient home program were well in place and that programs soon to be established included weatherization, implementation of mandatory curtailment if and when it occurred, customer consumption comparison data, a home energy analysis team (HEAT), an energy-audit program and an extensive public relations program. Scofield said that future activities would include utilization of an EWEB in-house conservation team, a mobile exhibit, homeowner-energy seminars, an advertising program and the monitoring of legislation. EWEB's conservation efforts were well-received by ratepayer Ray Wolfe and Scofield was praised for his efforts by Commissioner Camilla Pratt who had been elected to the Board in November, 1976.

An important step was taken in October 1977 when the Board approved participation in the Energy and Man's Environment program (EME). This activity, a cooperative effort of both investor and consumer-owned utilities in the Pacific Northwest, was designed to objectively take to the public schools comprehensive information about energy and its relationship to the environment. This was a sharp departure from an earlier school program "This Atomic World" which, while very professionally conceived and implemented, presented a biased picture of nuclear power and its importance. On the basis of General Manager Parks' explanation of the EME program, the Board approved $18,500 for its participation on October 24, 1977.

In November EWEB Rate Analyst Garry Kunkel addressed the Board on the rate philosophies included in President Carter's National Energy Act then under discussion in Congress. Kunkel told the Board that the proposed legislation included such rate mechanisms as time-of-use pricing, seasonal rates, the elimination of declining block rates and master metering, monetary rewards for load management and lifeline rates. Lifeline rates, which were to become controversial in the months to come, represented a concept in which consumption of electricity to meet the basic needs of a customer for the maintenance of health and comfort would be at a reduced price, and in this sense, the lifeline rate was a form of the inverted rate previously discussed. Just two days after Kunkel gave his presentation to the Board, Congressman Jim Weaver appeared at a special Board meeting to present his own version of a national energy act. Weaver preferred the term "baseline rate" to define what others were calling the lifeline rate. Weaver was told by the Commissioners and General Manager Parks that the baseline rate would be extremely difficult to administer and would also be, in many instances, inequitable in application.

In November, 1977, the city of Portland filed suit in order to acquire preference power for residential customers within that city. General Manager Parks requested and was granted authority by the Board to enter into such litigation as was necessary to fight the city of Portland's effort. The Board also earlier in the year had gone on record opposing Senate Bill 320, which was the DRPA legislation proposed by Governor Straub.

In his annual message on January 9, 1978, President Tiffany alluded to continuing accomplishment in such areas as conservation, the water system improvement program, a good working relationship with the 1977 Legislature and local citizen participation in EWEB affairs. He recommended an even stronger conservation program. But perhaps the most important objective he stated was the development of a viable regional power program.

Commissioner Taylor Ramsey had resigned effective at the end of 1977, and at a special meeting on January 25, 1978 Thomas L. Teutsch was appointed.

In February 1978 it was announced that total in lieu of tax payments to local governments would amount to over $3.7 million. This was the largest amount to date and it included $1.9 million resulting from

EWEB's Trojan Nuclear Plant participation. In March the new Rams Realty building was dedicated as the first commercial energy-efficient project in the Eugene system. This was another example of EWEB's early and timely efforts in the area of conservation.

EWEB continued its close attention to wages paid in the utility industry and in April the Board approved a general increase of 6.8%. An examination of fringe benefits revealed that EWEB was then well in line with practices in the industry and no changes were made.

ACCOMPLISHMENTS LISTED

By summer, 1978 EWEB was in a position to enunciate accomplishments that had been realized in the conservation field. A weatherization program had been approved by the Oregon Department of Energy, administrative rules for mandatory curtailment had been adopted by the Oregon Public Utility Commission and the EWEB implementation plan for mandatory curtailment had been completed. Twelve hundred homes in Eugene had been audited and consultation had been provided to 300 large commercial and industrial customers. In addition, over 100 speaking engagements had been filled and a pilot energy-education program for 150 pre-school chidren was underway.

In August, 1978 Rockne Gill, senior attorney for Schwabe, Williamson, Wyatt, Moore and Roberts, explained the Pacific Northwest Power Planning and Conservation Act to the Board. Gill addressed provisions in the current version which were of concern to the Public Power Council. In order to overcome such concerns it would be necessary to seek amendments which would assure continuing construction already underway before turning to new resources—such construction delays being very costly to ratepayers, preclude Bonneville from building new generating plants, protect the preference clause, provide a broader, more workable definition of "cost-effective" as applied to resources, and provide a better definition of average system cost. Gill added that the current version of the legislation, if the Public Power Council concerns and amendments were taken into consideration, probably would be the best option from the standpoint of EWEB.

In October 1978 Commissioner Camilla Pratt reversed her position on the concept of the lifeline rate, or "baseline rate" by Congressman Weaver's definition. The Board was to remain from that point on solid in its opposition to this controversial rate design.

As it adopted the 1979 budget, the Board in December, 1978 also approved a 4½ percent electric and steam rate increase and a seven percent water rate increase. By this time the Board was well aware that the electric system was subsidizing the steam system in terms of return on investment. The attendant problem was that if steam rates were advanced to a level commensurate with the cost of providing steam heating service, steam customers would be given a strong incentive to convert to electricity for heating and other industrial purposes. This would be in contradiction to the Board's desire to hold down the increase in electrical consumption. Raising rates for steam service to a full cost-of-service basis might also adversely affect the role that EWEB had taken under General Manager Byron Price in improving air quality in the Eugene area through the efficient burning of wood waste in the EWEB system. To the extent that this was a dilemma to the Board, it was a dilemma based upon the fact that a situation had evolved in which there were no precise, easy answers. The Board was making the best decisions that it could under existing circumstances.

FORMAL REGULATION EMERGES

The National Energy Act had been passed by Congress on November 9, 1978. It was to bring to EWEB a formal rate regulatory process for the first time in its history. Some explanation of this point is in order.

In a very real sense, EWEB had been *regulated* from its beginnings. This was regulation in the form of control by the Board which represented the citizen constituency. However, regulation had not existed in the formal sense which included such mechanisms as evidentiary hearings.

Ralph Carl Louis Hoeber, in his scholarly doctoral dissertation published in 1948 entitled *The Development of Public Utility Regulation in Oregon,* provides an excellent chronological history of regulation in the state. Prior to 1887 regulation had been limited to some statutory control of ferrymen, millers, corporate charters and franchises. A Board of Railroad Commissioners was established in 1887 and while deemed generally ineffective, served as something of a model for future regulatory departments. Modern regulation began with the Public Utilities Act of 1907. The Act included the following definition:

> The term "public utility" as used herein shall mean and embrace all corporations, companies, individuals, associations of individuals, their lessees, trustees or receivers (appointed by any court whatsoever), that now or hereafter may own, operate, manage or control any plant or equipment or part of plant or equipment in this state for the conveyance of telegraph or telephone messages with or without wires or for the transportation of persons or property by street railroad as common carriers, or for production, transmission, delivery or furnishing of heat, light, water or power; and any and all whether said plant or equipment or part thereof is wholly within any town or city, or not.
>
> No plant owned or operated by a municipality shall be deemed a public utility under or for purposes of this Act.

Explicit in this definition was the political philosophy, particularly strong in Oregon, that local control should be preserved wherever possible. By 1913 municipalities which were acting informally as their own regulatory bodies were serving nearly 22,000 customers. These municipalities included Ashland, Canby, Canyonville, Cascade Locks, Drain, Eugene, Forest Grove, McMinnville, Milton and Monmouth.

The Oregon Railroad Commission was renamed the Public Service Commission in 1915. In 1931 the agency was renamed again the Oregon Public Utility Commission and essentially the organization has remained intact since. Commission structure in Oregon had evolved toward a single commissioner concept. A several commissioner concept was criticized because it might "lack energy and promptness of decision," have difficulty in placing responsibility, and tend to become conservative, bureaucratic and unresponsive to "the voice of the people."

Whatever the composition of the Oregon regulatory body, the Eugene municipal system escaped formal regulation from higher authority until the National Energy Act of 1978, and even with the federal legislation, regulation, while becoming vastly more proceduralized, would essentially remain in the hands of the local Board of Commissioners.

In February 1979 the Board heard a presentation on the impact of the National Energy Act, particularly those sections of the Act known as the Public Utilities Rate Policies Act (PURPA) and the National Energy Conservation Policies Act (NECPA). The presentation was given by Garry Kunkel, John Scofield and Jean Reeder, the latter having been in the employ of EWEB for a relatively short time. In March Jay Waldron of the firm of Schwabe, Williamson, Wyatt, Moore and Roberts outlined the hearings requirements of the National Energy Act. In May Waldron and Kunkel discussed with the Board hearings procedures, and the Board adopted a manual for hearings under the National Energy Act. At a later meeting in May, Orlando John Hollis was appointed presiding officer and at yet another special meeting in May, a schedule for PURPA hearings was adopted.

HEARINGS SUCCESSFUL

The hearings began on September 17, 1979 and were subsequently continued on September 24 and 25, October 8 and November 2, and were held at the EWEB Tour & Education Center. These hearings were essentially designed to establish ratemaking procedures as well as taking into account certain aspects of rate design. Orlando John Hollis, as presiding officer, conducted the hearings in an expeditious, orderly and highly intelligent manner and once again made a valuable contribution to EWEB and his community.

The procedure followed in the hearings was in accordance with the provisions of the manual prepared by Jay Waldron and adopted by the Board. Although copies of the manual were available to all interested persons prior to the hearings, no person filed as an intervenor as provided for in the manual. Waldron, in his capacity as special counsel to the Board, represented the staff throughout the hearings.

Mel Damewood, John Fuller, Steve Hill, Garry Kunkel, Keith Parks, EWEB rate consultant Dan Reed, Tom Santee, Don Vanderzanden and Harold Worcester appeared as witnesses on behalf of EWEB. Private citizens appearing were Chris Attneave, Bruce R. Berg, John H. Van Landingham IV, Jon Silvermoon and David Tobin. Two members of the Oregon Public Utility Commissioner's staff who appeared were Wallace Gibson and Evan White.

EWEB was included under PURPA because its annual sales of electricity exceeded two billion kilowatt-hours per year. The Eugene utility operation was included in the official list of covered utilities which was issued by the Secretary of Energy.

EWEB rate analyst Garry Kunkel, working closely with Jay Waldron, played a major role in the organization and ultimate success of the hearings. The essential function of the hearings was to examine the appropriateness of the various standards set forth in PURPA and their applicability to EWEB—standards relating to cost of service-based rates, declining-block rates, time-of-day rates, seasonal rates, interruptible rates, load-management techniques, master metering, automatic-adjustment clauses, information to customers, procedures for termination of electric service, advertising and lifeline rates.

The hearings supported several levels of appropriateness that were the basis for Board decision as to adoption of standards. Standards for cost of service, declining-block rates, interruptible rates and load-management techniques were adopted without qualification. The standard for time-of-day rates was not adopted but the Board was to order that this form of rate design be studied. The seasonal rate standard was not adopted nor was the concept of the lifeline rate.

During testimony it was pointed out that lifeline rates as implemented in the state of California were not deemed successful by many experts in the field. One of the problems cited was that as a practical matter, if lifeline rates are in place, they are available to all customers and not limited to the needy and elderly. Another problem in California was the tremendous shift of the revenue requirement burden to commercial customers who did not benefit from the lifeline rate.

The Board would find that the elimination of master metering was appropriate, but only within the context of EWEB's existing practice, which was to individually meter all customers living in multi-unit residential buildings, and did not adopt the standard pertaining to the elimination of master metering in commercial and industrial buildings.

Automatic-adjustment clauses were deemed inappropriate. Fuel adjustment clauses, as they had existed in such areas as southern California, were based on the concept that there could be rapid changes in the price of oil which was used for generating purposes. The adjustment clause was a means of automatically taking care of such fluctuations in the price of the fuel used for generation. There did not appear to be a need for such an adjustment clause in a regional system which still had a substantial hydroelectric base without attendant rapid fluctuation in the cost of generation.

The standard dealing with information to customers was generally adopted in principle but with several departures from the suggested standard which were in keeping with current EWEB practice. The standard on advertising was determined to be inappropriate for adoption but as an alternative the Board determined that the unwritten advertising policy which had been followed for many years would be formalized and that this would adequately meet the objective of the standard.

The EWEB response to the Public Utility Rate Regulatory Act was an example of excellent administration and organization. General Manager Keith Parks, Garry Kunkel, Jay Waldron and others worked as a team in carrying the hearings through to a successful conclusion which culminated in a final order by the EWEB Board of Commissioners on April 7, 1980. Orlando John Hollis, as presiding officer, conducted the hearings with his inimitable mixture of courtesy and sternness. He did not permit departure from the point at hand and at the same time he treated all with fairness and respect.

CLASH OVER REGIONAL ACT

Another area of federal legislation came to a Board meeting on September 10, 1979 at which Deputy Public Utility Commissioner Roy Hemmingway appeared to seek Board support for S 885, which was Senator Henry Jackson's version of a Regional Energy Act. The Board had heard a lengthy pro and con discussion of the Act by special counsel Larry Cable in August.

Following Hemmingway's request for Board support of the Regional Energy Act, a heated discussion ensued in which General Manager Parks and Board members questioned the effect the legislation would have upon the preference clause which had benefited Eugene and other consumer-owned systems for many years. It was also asserted by Board members that the legislation was political in nature, designed to "bail out" the investor-owned utilities in the region through an exchange agreement in which the utilities could acquire preference power for their residential customers, thereby reducing the amount of such power available to the consumer-owned systems.

Hemmingway did not obtain what he sought. President Tiffany closed the discussion by calling for a vote on a resolution opposing Senator Jackson's bill. The resolution was adopted with a vote of 5-0.

ATTEMPT MADE TO UNSEAT TIFFANY

Newly elected commissioners Jack Craig and John Bartels urged adoption of rules of order at the organizational meeting of the Board which was held in January, 1979. The staff formulated such rules and they were adopted on February 12, 1979 along with bylaws for the EWEB Board. The two new Commissioners also attempted to unseat President John Tiffany at the 1979 organization meeting when they brought about the nomination of Camilla Pratt for Board presidency. Pratt declined to be a candidate for the position and instead nominated John Tiffany, who was reelected to the position by a vote of 3-2.

In early 1979 the Board approved in principle and heard plans to institute a customer response program designed to give all customers the opportunity to make comments or ask questions regarding the utility operation and its services. This concept was a contribution of President John Tiffany, who had employed a form of customer response program in his retail drug store operation. The development of the customer response activity was to involve a task force of several EWEB employees with Garry Kunkel taking a leadership role.

The customer response program which eventually was implemented proved to be a very valuable tool in measuring trends of customer opinion and determining areas in customer relations where problems might exist. The program was designed to assure that customers would receive prompt answers to their questions. The program would become even more significant when escalating rates for both water and electricity would bring an ever-increasing number of high-bill inquiries and the need for explanation.

The task force method of approaching problems had become a basic tool in Keith Parks' mode of administration. The concept was a simple one. It involved bringing those personnel together who were closest to the problem at hand and giving them the latitude to thoroughly explore the problem and develop recommendations. The customer response program was a notable example of this process.

In April 1979 the Board granted a seven percent interim wage increase. This was later revised to 9.4 percent when more complete information on Pacific Northwest utility settlements was obtained. No benefits were added, but at the previous meeting of the Board the Veterans Day holiday was changed to the day following Thanksgiving. This action was based upon a trend in the industry and a consensus of EWEB employees.

Dick Anderson had joined the organization on April 1, 1978. He came from the Las Vegas, Nevada municipal water system and brought with him a great amount of professionalism and expertise in the area of labor relations. He quickly established good working relationships with the other utilities in the region and it appeared that the maintenance of positive labor relations and personnel management in the Eugene municipal system was in good hands.

THREE MILE ISLAND DISCUSSED

The Three Mile Island nuclear plant accident which occurred March 28, 1979, prompted lengthy discussion among Board members, particularly Camilla Pratt, John Bartels and Jack Craig. Chris Attneave, who was present at the meeting on April 2, participated in this discussion. Three Mile Island would profoundly influence the nuclear generation industry in the United States. The event gave credence to those who contended that nuclear power was unreliable and dangerous. Industry spokesmen, on the other hand, adopted the point of view that Three Mile Island demonstrated that safety systems worked. In any event, Three Mile Island, coupled with incredibly escalating costs for nuclear plant construction, was a serious blow to the nuclear generation industry in the United States.

In April 1979 EWEB staff members John Scofield, Jack Clark and Norm Stone went to Washington, D.C. to work as members of an American Public Power Association committee dealing with proposed rules and regulations for the National Energy Conservation Policy Act, which was the other component of President Carter's National Energy Act which directly affected the Eugene utility system. It was nearly a year later, in March, 1980, that Federal Legislation Manager John Scofield outlined to the Board the provisions of the National Energy Conservation Policy Act and the Residential Conservation Service. At a later meeting in March, hearings procedures for implementation of this Act were announced. Once again Orlando John Hollis would serve as presiding officer for the hearings which were to be held April 14-18, 1980.

Twenty-three witnesses registered to testify during the hearings. Jean Reeder, who had been appointed conservation manager on February 1, 1980, Federal Legislation Manager John Scofield and Community Information Officer Tom Santee were scheduled to present exhibits on behalf of the staff. City Councilor Gretchen Miller would present testimony on behalf of the Eugene City Council's Energy Conservation Policy Board, of which she was chairperson. A considerable amount of written testimony also would be received. During the hearings Tom Santee outlined the procedure which had been followed for public notice of the hearings. John Scofield and Jean Reeder shared in the presentation of the various components of the plan, notable of which were definition of the scope of benefits; the program announcement as required by the federal legislation; energy analysis; lists of suppliers, installers and lenders; arranging installation; arranging financing; post-installation inspections; and establishment of qualifications for energy analysts, installers and inspectors.

Other components included complaint procedures, reporting and record keeping, and accounting and payment of costs. Much of this testimony, both oral and written, dealt with the relationships between the Residential Conservation Service program and contractors and installers. Questions were raised regarding such matters as the effect upon EWEB should a contractor with outstanding warranteed work go out of business or refuse to perform. Another question raised was whether or not contractors approved under the program should be required to belong to a local professional organization. City Councilor Gretchen Miller urged that the financing program be equitable to all in the community, including those who have already weatherized their homes. She also suggested further studies by the EWEB staff in order to determine the true avoided cost of the construction of new generation as a means of measuring the cost effectiveness of conservation.

Again, as in the case of the PURPA hearings, the NECPA hearings were conducted in an excellent manner under the supervision of presiding officer Orlando John Hollis and with the coordination of John Scofield and Jean Reeder. On May 5, 1980 Hollis, Reeder and Scofield reported on the NECPA hearings and a resolution approving EWEB's Residential Conservation Service plan was adopted on May 27. The PURPA rate hearing process was put into operation in late summer, 1980 and proved to be a workable vehicle to assure not only compliance with federal law but also establishment of an ongoing systematic means of providing public input in the ratemaking process.

Other significant events had occurred during 1979. In June the Board approved a grant application to the U.S. Department of Energy for a $300,000 geothermal research grant. The work was to be conducted in the Cascade mountains in the state of Oregon. By early July the grant was approved and planning for the digging of test holes was begun.

The Board also took action in 1979 to enter into an agreement with the Bonneville Power Administration to participate in an infrared fly-over program designed to provide graphic demonstration of the effects of ceiling insulation on heat loss from roofs of houses. This program would eventually provide some very interesting aerial photography of residential areas in the Eugene community. However, problems developed in utilizing this technology effectively in a meaningful public information program and national interest in infrared fly-over activity eventually waned.

Minor conflict occurred at the Board meeting on October 1, 1979 when President Tiffany raised the issue of Commissioners Craig and Bartels using EWEB letterhead to express their individual views on various topics. This question was resolved when General Manager Parks was directed to provide individual letterheads and business cards to commissioners upon request.

CONTROVERSIAL DECISION

One of the most controversial and misunderstood actions ever taken by the Board was its unanimous decision on January 7, 1980 to leave its share of Trojan power net billed through the Bonneville Power Administration. The immediate reaction in the business community was "Why did EWEB sell its share of Trojan?"

Net billing was conceived in the Pacific Northwest as a means of protection against a "dry hole" event in which a utility or group of utilities might invest a vast amount of money in a construction project only to have it fail because of unforeseen events. By agreeing to assign the output of such a project to Bonneville, the federal agency would in turn bear the cost of the project. If the project succeeded, Bonneville would add the output to its power pool and sell it to the utilities throughout the region at a melded hydrothermal rate. If the project were to fail, Bonneville, under a net-billing contract, would absorb the cost of that failure. In the event of either success or failure, gains or losses through the net-billing arrangement are in effect distributed to utilities and direct-service industries throughout the region.

In 1976 BPA had issued its letter of insufficiency in which it stated that it would not be able to meet all preference customer demands after July 1, 1983. This notice was later to be rendered invalid through regional legislation, but nevertheless, it was a factor for consideration in January 1980.

EWEB's consulting engineer, Ford, Bacon & Davis, developed recommendations relative to possible Board action on whether to withdraw all or part of its share of Trojan power or to leave it net billed with BPA as had been the case since construction of the Trojan project had begun. If Trojan power were to be taken into the Eugene system, the engineers' report indicated that this would not be necessary before 1991. At that point the principal advantage of Trojan would be its relatively low-cost power as compared to other alternatives available to the Board. The engineer estimated that in 1991 the cost of Trojan power at a 35 percent plant factor would be 70 mills per kilowatt-hour as opposed to 123 mills for geothermal power and between 100 and 132 mills for wood waste generation. By the year 2011 Trojan power was estimated to cost 177 mills at 35 percent plant factor as opposed to 393 mills for geothermal and 322 to 424 mills for wood-waste generated power.

Bob Hazeldine, representing Ford, Bacon & Davis, also spoke to the superior management and operating condition of the plant as an advantage in withdrawing Trojan power and that the plant "holds promise for giving satisfactory performance during its remaining life."

In addressing disadvantages of taking Trojan power, the engineer pointed out that all economic advantages would dissipate immediately if the plant were to shut down. It was also pointed out that efforts to shut down all nuclear plants might succeed. Perhaps most importantly, the engineer cited the Three Mile Island incident as being of critical importance in evaluating the feasibility of Eugene's depending upon Trojan power.

The engineer concluded, "The decision cannot be made on the basis of empirical data alone; it also requires a judgment of the risks inherent in all choices."

Hence the decision was one which was very complex. However, it would be an error to say that the decision was difficult for the Board, inasmuch as in Eugene and on the Board itself the temper of thought was sharply in opposition to nuclear power.

The frustration of the businessman on the street could be answered by saying that it was a Catch-22 dilemma that the Board faced. If the Board elected to withdraw Trojan power and for whatever reason, the plant was permanently shut down, the Eugene utility and its ratepayers would be held to the payment of debt service and the costs of closure and the amount of money involved would be vast. On the other hand, the decision to leave Eugene's share of Trojan net billed with the Bonneville Power Administration would relieve Eugene of future financial burden including the ultimate decommissioning of the plant. It could also be said that if a plant has good operating experience throughout its life and its relatively low-cost power were to be melded into the Bonneville wholesale rate, Eugene, along with all other Bonneville customers, would benefit.

The decision was a gamble. Whether it will cost or save Eugene ratepayers vast sums of money will not be known until the turn of the century.

FEDERAL LEGISLATION COSTLY

In August, 1980 the Board was informed that over half a million dollars had been spent by the utility as a result of the passage of the National Energy Act and another $462,000 was approved in October for the purchase of special metering equipment to meet the load management research requirements under the Public Utility Rate Policies Act.

"FULL-EMPLOYMENT ACT" FOR ATTORNEYS

The Pacific Northwest Power Planning and Conservation Act was signed by President Carter on December 5, 1980. Alan Richardson, legal counsel for the American Public Power Association, had said shortly after passage of the National Energy Act that the Act "could be called a full-employment act for attorneys." Richardson's jocular observation had already begun to have substance and the regional legislation would eventually support a similar observation.

A hard-working Board of Commissioners, while devoting much time and energy to regional and to some extent national afffairs, also was confronted with local problems. Public hearings were held during 1980 on the Jefferson-Charnelton Substation and accompanying 115 kV transmission line. This was to result in a location of the transmission line along 13th Avenue which, while politically most acceptable, was not deemed the optimum choice by some members of the staff. But as the Board had grown to understand, decision making could not always be accomplished on the basis of precise technical information. There were issues of consensus to be considered. The role of neighborhood groups had strengthened considerably in recent times. Often the Board found it necessary to weigh broad community goals against the concerns of perhaps a relatively few number of citizens who might object to a particular concept that the Board was considering.

Early in 1981 the Gillespie Butte property, which had been acquired as a reservoir site in the 1950s, was declared surplus and offered for sale. This stimulated considerable controversy. Some suggested that EWEB should give the property to the city for park purposes. Another suggestion was that the city owned the property anyway, inasmuch as all EWEB property was owned by the city of Eugene on behalf of the Eugene Water & Electric Board. These arguments were turned aside and the property was placed on the market at an asking price of $500,000.

MANDATORY WEATHERIZATION ENDORSED

In February 1981 the Board unanimously endorsed a mandatory weatherization ordinance which was being considered by the City Council. The ordinance would require, by 1985, that before a residence could be sold it must be weatherized. This ordinance demonstrated the active interest that local government apart from EWEB was taking in energy matters and conservation.

DISCUSSION CONTINUES

The ongoing dialogue between the Board and the City Council gained some emphasis during this time when Commissioners Camilla Pratt and Sarah Hendrickson were appointed to serve with Councilors Betty Smith and Emily Schue as a City-EWEB task force. At an earlier meeting Councilor Brian Obie and City Manager Mike Gleason had appeared to discuss some problems that existed between the two bodies. At a later meeting the Board generally approved the concept of General Manager Keith Parks and City Manager Mike Gleason working out mutual problems.

The Board had taken the position earlier that the matter of the Gillespie Butte property as well as the property owned by EWEB adjacent to the city hall was City Council business and should be deferred to that body. The city would act in the near future to acquire the downtown property which had been purchased earlier by EWEB as a possible office building site.

In May, 1981 the Board departed somewhat from its practice of several years of granting a wage increase generally across the board with nearly all employees being affected. The Board had adhered to the earlier practice in 1980 when a 13.5 percent wage increase was granted. However, the 1981 increase of 11.4 percent applied only through the middle management ranks of the organization and above that was graduated downward to six percent. No changes were made in fringe benefits and this was in part due to the fact that EWEB agreed to pay the employee's contribution to the Public Employes Retirement System effective January 1, 1980. This added six percent to the employer's fringe benefit costs and, accordingly, EWEB was well within the upper ranges of fringe-benefit programs among Northwest utilities.

In June the Board approved limited participation in the Creston coal-fired plant which was being planned by Washington Water Power Company. This overture toward conventional thermal generation was the first that the Board had made since the abandonment of a Eugene-sponsored nuclear plant in 1974.

DELAYS ENCOUNTERED

There was a great amount of difficulty experienced in implementing the Residential Conservation Service program under the National Energy Conservation Policy Act. Delays in obtaining conservation contracts from BPA prompted the Board to instruct General Manager Parks to write Senator Mark O. Hatfield about the problems the utility was encountering.

WPPSS 4 AND 5 ROCK REGION

Early in the summer of 1981 the region was rocked by the news that Washington Public Power Supply System Plants 4 and 5 were in a desperate financial situation. In 1974 the Board had wisely declined to participate in these two plants, but it was soon apparent that WPPSS 4 and 5 could very well have long-term adverse effects upon all of the utilities of the region, including those which were not participants.

Investigations of small hydro projects continued and in the fall of 1981 the Board heard presentations regarding Dorena, Blue River, Minto and Sunnyside as possible sites for construction. An application for the Blue River project would be approved by the Board one year later.

In September, 1979 General Manager Parks had told the Board of the possibility of EWEB becoming involved in a geothermal pilot project in Nevada. He said that EWEB could realize as much as 10 megawatts of the project's output. Herb Hunt, in December, 1981 presented an overview of the project. The Board adopted a resolution establishing EWEB as a member of the Nornev Demonstration Geothermal Company (NORNEV). Other members of the enterprise were Sierra Pacific Power Company, Sacramento Municipal Utility District and Pacific Power & Light Company. Hunt served as principal EWEB representative for the project.

As 1981 closed, elements of the National Energy Conservation Policy Act began to be operational and the Board heard a presentation on the energy-efficient street lighting program. On November 30, Conservation Manager Jean Reeder explained the energy buy-back program and how it was working in Eugene. "Energy buy-back" meant that the Bonneville Power Administration bought back from utility consumers the energy saved over a period of a year through a conservation effort on the part of the consumer. Such energy buy-back was justified by the "avoided cost" or the cost of new generation which could be avoided because of a regionwide conservation effort. The payment for buy-back was set at 29.2¢ per kilowatt hour. This was deemed to be cost-effective in terms of what new generation would cost. This figure would hold for some time except that at the urging of the Regional Council, which would be established by the Regional Act, the buy-back figure for low-income consumers undertaking weatherization was raised to 37.8¢ per kilowatt hour effective September 15, 1984.

RELATIONSHIPS COMPLEX

The relationship of utilities to the Bonneville Power Administration, their relationship to the Pacific Northwest Electric Power Planning and Conservation Act (Regional Act), and to the Regional Council which was formed as a result of that Act, were becoming major issues in the Northwest. Also in question was the relationship between the Regional Council and BPA. It appeared that a major political struggle was in process and that utilities would have to survive as best they could in an environment of confusion and complexity.

Libby Henry, EWEB's capable legislative representative, gave a report to the Board in April, 1982 on the BPA billing credit concept. She said that billing credits, then in the rule making process, were unilateral and nonnegotiable and that the concerns she would set forth were those of the Public Power Council.

As stated earlier, the letter of insufficiency had been negated by passage of the Regional Act which in turn provided that BPA was obligated to meet the loads of its customers and to insure an adequate supply of power to meet such loads. Billing credits were one means of meeting this requirement.

Billing credits, simply stated, meant that BPA would credit its customers' power bills for conservation or certain resource development undertaken by such customers—such credit being based upon the alternate, or avoided, cost which would be involved if BPA acquired new generation.

Beyond this simple concept, billing credits became exceedingly complex and of doubtful incentive value toward encouraging action on the part of the utilities. Among other concerns, Henry spoke to the imposition of negative billing credits which could be applied to resources other than conservation. This would require that a customer receiving a billing credit begin paying BPA at such time as the customer's BPA power rate exceeded the net cost of the resource for which the credit was granted or the alternate (avoided) cost of new generation which would have had to be acquired by BPA.

After Henry presented her analysis the Board supported the Public Power Council with Commissioner Pratt suggesting that "a strongly-worded statement from the Board supporting the position of the Public Power Council be sent to Bonneville and the Regional Council."

The billing credit issue exemplified the growing complexity of energy affairs in the region. Fears that the Board had expressed earlier about the Regional Energy Act were being borne out and already the relationship of the Regional Council to Bonneville Power Administration was unclear and confusing to the energy industry in the Pacific Northwest. Roy Hemmingway, who had made a passionate entreaty to the Board in September, 1979 for its support of the Regional Act, had been appointed by Governor Vic Atiyeh as one of two Oregon representatives to the Regional Council.

Thousands of school children benefited from classroom presentations and tours conducted by William E. Eaton; this work continues through efforts of Al Hughes, Charlie Strange and Evan Gentry. (EWEB files)

WAGE FREEZE IMPOSED

In May, 1982 the Board approved a wage freeze for the entire work force. This holding of wages at present levels was imposed out of respect for the severe economic conditions in the Eugene area. Save for a few instances, there appeared to be no severe employee objection to the freeze even though substantial increases had been granted to other utilities in the region. The general lack of reaction was a tribute to EWEB employees and their undertanding of the local economy.

As 1982 drew to a close, it was becoming increasingly evident that the Regional Energy Act and the Regional Council which it established added yet another layer to the energy bureaucracy of the Pacific Northwest. In a report to the Board on November 1, Rate Manager Garry Kunkel said that the council was moving ahead without adequate recognition of the Public Utility Rate Policies Act. Many of the assumptions employed by the council were unsubstantiated. Among these were positions being taken by the council on such matters as rate impact upon various income groups and the relationship between energy and capacity planning. Keith Parks told the Board that the council was infringing upon the rights of the local governing body. The Board in turn encouraged Parks to fight this trend in any way that might be possible.

In his annual message on January 3, 1983, President Tiffany said that the only electric rate increase which might be imposed during the coming year would be a BPA wholesale rate increase pass-through. He also emphasized the importance of customer sovereignty, pointing out that while it had once been in the customer interest to promote consumption of electricity, it was now in the best interest of customers and the electric system to encourage conservation. Tiffany also underscored the necessity of support for the general manager and his staff and to maintain high productivity and fiscal integrity. He also emphasized the need for additional research and development and the desirability of EWEB participating in community affairs, particularly the planning process.

In a reference to the famous novel by George Orwell, Vice President Camilla Pratt said that "we are uncomfortably close to 1984." She was obviously referring to the barrage of new laws, regulations and bureaucracies which were affecting EWEB with ever-increasing impact.

Chief Engineer Owen Brown presented the subject of the Willakenzie-Spring Creek power transmission line project. He proposed a timetable of Board presentations and hearings which hopefully would lead to the beginning of construction by March 30, 1984. Four hearings were held during the summer and there was a vast amount of controversy and public input regarding an EWEB alternative site which would have provided for a river crossing 1½ miles downstream from the Beltline Bridge. The route, which was finally adopted by the Board on June 18, 1984, provided for a crossing of the river at Beltline Bridge along the path of an existing power line. The line was estimated to cost $2.3 million.

The Willakenzie Spring-Creek transmission line, needed to enhance electrical system reliability in north Eugene and the River Road area, represented yet another example of significant public involvement in the decision-making process. The EWEB engineering department, using a sophisticated analysis process, had developed many alternative sites for the line.

The decision had been delayed by several months because of the controversy and the concerns of the Board. But as Commissioner Camilla Pratt observed at a meeting during 1983, better decisions will be made if they are based upon constructive public dialogue.

In January 1983 the Board approved a resolution selling its downtown property to the city of Eugene at a price of $950,000. Board interest had waned in a downtown location for an EWEB headquarters building in recent years and the city felt that it needed the property for the expansion of its own facilities. In July, Chief Civil Engineer Jim Brown told the Board that requests for proposals were being submitted to several architectural and engineering consulting firms toward the objective of developing a master plan for the EWEB headquarters location.

The 1983 Legislature was given a great amount of attention by the Board. House Bill 2913, also referred to as the Bennett Bill, would pro-

General Manager Parks continued high-priority emphasis on industrial safety. Elliott Lewis, 18-year veteran with the municipal-utility operation, took over safety-manager job in November, 1982. (EWEB files)

hibit any hydroelectric construction or maintenance on that portion of the Santiam River where EWEB's proposed Minto Dam site was located. This bill, when it was passed by the Legislature, had a total of 53 sponsors. The significance of this particular legislation lay in the precedent that it established and the potential that any individual project pursued by any entity within the state of Oregon could be eliminated by a specific legislative act. This does not mean that the Legislature has not had such authority through the years. However, Representative Chuck Bennett demonstrated that, given the appropriate political circumstance, a project could be put out of business by legislative action.

Also of concern to the Board were the perennial legislative bills pertaining to public contracting. In nearly every session an effort is made to force public agencies to take their public works activities to the private sector. Some limitation on the amount of work a public agency can perform has been in effect for many years. As the 1983 Legislature drew to a close, only one significant public contracting bill was passed. Senate Bill 190 abolished the Public Contract Review Board and transferred the functions involved to the Oregon State Department of General Services, with this department's actions affecting state agencies only. The bill also extended to school districts the authority to form their own public contract review boards. Inasmuch as the EWEB Board has acted for several years and continues to act as its own contract review board, the legislation had little effect upon the Eugene utility operation.

Legislative representation in Salem had been formalized at EWEB in 1970 when former state senator and radio station owner Glen Stadler was hired. His established tradition of an excellent representation at the Capitol was continued by Libby Henry. EWEB's lobbying activity has been essentially characterized by intelligent and informed relationships between EWEB and the legislative body. This has consisted in many instances of providing full and accurate information to legislators, including the consequences of legislation which has the potential of adverse effect upon the state and its communities.

In April 1983, as had happened in earlier times, the Board was questioned about workers being brought into the area to perform work for the municipal utility. Several individuals appeared at a Board meeting, complaining that the contractor, Siemens-Allis, Inc., had brought workmen in from California to perform work at the Carmen-Smith power plant. Both President Tiffany and Keith Parks responded that the hiring of employees is the province of the contractor and that EWEB had no control over this process. However, as a result of this demonstration of interest on the part of local citizens, the contract language used in bidding work out to contractors was strengthened to the degree possible to encourage the employment of local people.

In June Commissioner Camilla Pratt reported that General Manager Keith Parks had been awarded the Distinguished Service Award by the

Northwest Public Power Association at its annual meeting in Kalispell, Montana. Commissioner Pratt said that she "could not think of anybody in the entire Northwest over the last five or ten years who could possibly have deserved it more and we should be very proud of him because it is a fine honor." Commissioner Pratt's remarks regarding the general manager were representative of the Board's general attitude toward its chief executive officer.

On July 11, 1983, Conservation Manager Jean Reeder introduced to the Board the concept of revenue-bond financing for conservation purposes. She said that such bond financing could be an alternative to a long-term conservation contract with BPA. On August 1, Reeder and Assistant Treasurer Don Vanderzanden told the Board that if weatherization bonding could be achieved, debt service would be totally paid by BPA. This resulted in a resolution by the Board to go to the voters for a $30 million bond authorization. A task force was quickly formed to carry information about the upcoming election to the citizens of Eugene. Jean Reeder headed up the task force and other participants included Tom Santee, Deborrah Brewer, Shirley West, Norm Stone and Commissioner Camilla Pratt. Various kinds of information programs were developed and on September 20 the bonds were overwhelmingly authorized by Eugene voters. Later in the year the sale of the bonds would encounter difficulties in Congress along with other municipal securities offerings. But once again the voters on September 20 had given evidence of this confidence in the Eugene utility operation, and importantly, the Congressional problems affecting the bond sale were resolved in June, 1984.

In order to cope with the many problems that were arising with regard to energy matters in the region, General Manager Parks had formed "Task Force 1" in early 1982. Notable among the members of this group were—in addition to the general manager—Garry Kunkel, Jean Reeder, Hal Worcester, Dean Axtell, Vaughn Scales, Deborrah Brewer, Libby Henry and Staff Ecologist Dale Hagey. Work on various problems was divided among members of this group and specific assignments allocated. This task force became a very valuable tool for the general manager and the Board in not only keeping abreast of problems as they occurred but also in initiating action. EWEB in many respects was expending more than its share of effort in dealing with regional problems.

MAJOR REORGANIZATION REVEALED

During the fall of 1983 General Manager Parks revealed plans for a major reorganization of the utility operation. Three senior management members, Herbert H. Hunt, John E. Brown and Norm Stone, told Parks that they would retire at the end of 1983. Hunt originally planned to leave EWEB at the end of 1984, but revised his personal schedule and left a year earlier. He joined the city of Palo Alto utility department in February.

These retirements gave Parks the opportunity to be the architect of his own top staff. The new top echelon included Jean Reeder, age 40, who was appointed assistant general manager effective December 1, 1983. Reeder joined EWEB in June, 1978 as governmental affairs coordinator and in January, 1980 was promoted to the position of conservation manager. She holds a master's degree from the University of Oregon School of Community Service and Public Affairs. Garry Kunkel, also age 40, was appointed to the position of director of power management. He had over 19 years of experience with EWEB and had developed high capability in the area of rates. Rosemary Edwards, 55, for many years assistant secretary to the Board and with over 20 years of experience, was appointed director of administrative services, retaining the title of assistant secretary to the Board. Donald W. Vanderzanden, age 45, was appointed treasurer and director of accounting and finance. A graduate of the University of Idaho, he is also a certified public accountant. He had been with EWEB 15 years. In addition to the promotees, veteran Kenneth W. Rinard remained as director of engineering and operations. Chief Civil Engineer Jim Brown was promoted to the position of assistant director of engineering and operations. Dave Johnson was promoted to assistant treasurer and assistant director of accounting and finance. All of the promotions, except Reeder's, were effective January 1, 1984.

Other key employees remained in their positions and would contribute a great deal to the management hierarchy of the organization. Owen Brown would continue as chief engineer. Kimber Johnson remained as water operations superintendent. Dick Anderson would retain his post as personnel services manager and Charles P. "Pat" Mulloy would continue as purchasing manager.

Everett Jordan would remain as steam system superintendent and Steve Hill as credit and applications manager. Robert Carter would soon replace Dick Gall as electric operations superintendent and Greg Belshaw would remain as data processing manager. He would resign early in 1985 and be replaced by Sheila Mahan. Mat Northway would carry on as conservation manager.

In his annual message on January 9, 1984, President Tiffany commended the utility operation for its excellent performance during the recent Christmas weekend cold period. He also cited a report of EWEB's independent engineer, Howard Needles Tammen & Bergendoff, which gave very good marks to the performance of both the water and electric systems as well as the Trojan operation.

TIFFANY ANNOUNCES RETIREMENT

At the next meeting in January, President Tiffany announced that he would not be a candidate to represent Wards 1 and 8 and suggested that Jim Reinmuth, Dean of the School of Business Administration at the

University of Oregon, and who had indicated an interest in running, would be a good candidate.

For several months various staff members had been working on the design of the billing statement sent to EWEB customers. The Board had earlier indicated a keen interest in somehow providing greater clarification that the sewer charge, which EWEB collects for the city, was not an EWEB charge. It had been suggested at one point by Commissioner Craig that a separate billing be sent in order to eliminate confusion. On February 13 the Board adopted one of five alternative formats which were proposed. It provided a color-coded line for the sewer charge with adequate space for messages to customers and a comparison of electrical usage with the same period of the previous year.

In April General Manager Parks told the Board that the EWEB customer response program, during 1983, had prompted only about 6,000 questions and comments from EWEB customers, or less than one percent of the total 792,000 bills which had been mailed. This was an indication to the Board that the utility operation was in good condition.

PREFERENCE STATUS THREATENED

Regional problems grew even more complex and at the threat of the loss of preference status in the purchase of BPA power, the Board on April 23 adopted a resolution demanding "that BPA include in its final methodology the rate protection guaranteed by Congress which insures that BPA's preference customers will be no worse off with the Regional Act than without it." The same or similar resolution was adopted by governing boards throughout the region as a means of pressuring BPA not to disrupt the preference clause which had been in place for so many years.

By the end of June, 1984 EWEB had established an enviable record of conservation in its service area. Mat Northway, who had served as conservation manager since December 1, 1983, reported that over 35,000 energy audits had been performed on residences in the Eugene system. By the end of 1984 about 370 commercial establishments had had energy audits involving 1,600,000 square feet of commercial building space. Records indicated that 37 commercial establishments had undertaken aggressive conservation measures.

During the period 1973-1983 the average number of electric customers rose from 48,541 to 61,929, an increase of 27.5 percent. It should be noted that this increase occurred during the early years of the period when annual growth ranged between three and five percent. Increases during 1980, 1981, 1982 and 1983 were less than one percent. This related to Eugene's severe economic problems. The annual consumption of electricity in 1973 amounted to 1.7 billion kilowatt-hours while just over 2 billion kilowatt-hours were consumed in 1983—an increase during the latter period over the other of 17.6 percent, considerably less than the increase in average number of electric system customers.

Two factors contributed to decreased growth in electrical consumption. One is conservation. The other is price elasticity which means that when the price of a commodity rises, its consumption drops. Total annual electric sales during the period rose from $15,272,000 in 1973 to $57,218,000 in 1983, an increase of 275 percent. This reflects the high rate of inflation and the increased cost of purchased power.

The average number of water system customers during the same period increased from 24,309 to 33,476, a gain of 45.9 percent. The increase included acquisition of Oakway Water District customers in 1979. Water consumption during the period tended to remain constant. In several of the years consumption exceeded eight billion gallons but usage in 1983 was 12.5 percent less than in 1973. Price elasticity undoubtedly was a factor as were several damp summers.

Annual water sales increased from about $2 million in 1973 to slightly over $6 million in 1983, an increase of 208 percent. Heavy inflation was a major factor in this increase.

Notwithstanding the addition of a conservation staff of over 50 employees, the total work force increased only 13.6 percent between 1973 and 1983, from 380 employees to 432. This statistic is a tribute to the management of the organization.

At its regular meeting on Monday, October 22, 1984 General Manager-Secretary Keith Parks told the Board that he would retire at the end of the current year, stating that 39 years of public service were "about enough." Jean Reeder, who had been appointed assistant general manager-secretary December 1, 1983 would assume the position of acting general manager and so serve until the appointment of a permanent person to the post. (Following extensive nationwide recruitment, Reeder was appointed general manager-secretary of the Eugene Water & Electric Board effective July 1, 1985.)

If it is accurate that one of Parks' most difficult tasks from the beginning was the healing of old wounds left over from a bitter and tumultuous time of local controversy, he accomplished this mission well. Giving his attention to regional affairs, EWEB became the most articulate voice of all and had it not been for EWEB's remonstrance against the Regional Act and its ramifications, conditions today might be much worse than they are.

Parks' accomplishments have been set forth in foregoing pages. He can well be proud of his contributions to EWEB and to local government in Eugene and Lane County.

That remarkable French historian, Alexis de Tocqueville, said in 1848 that "society changes its forms, humanity its conditions, and that new destinies are impending." Dating from the time of the old Eugene Water Board, the governing body of the Eugene utility operation seems to have understood Tocqueville's admonishment. As trustee of a vitally important community service, it has served its constituency well.

CHAPTER EIGHT
Legal Counsel

WHEN THE EUGENE WATER BOARD was formed in 1911, it had need from time to time for legal counsel. For many years the Board utilized the services of the city attorney and records show that in the early days occasional payments were made to G. F. Skipworth for legal work rendered. Skipworth, who was to gain prominence as a Lane County jurist, was replaced by A. J. Foster, who served as city attorney and as needed, attorney for the Board until 1925 when he was replaced by S. M. Calkins.

S. M. Calkins served until 1946 at which time William East assumed the role of attorney for both the city and the Board. After one year, East relinquished his role as city attorney and the Board has since benefited from its own legal counsel independent of city hall.

In 1949, when East was appointed judge in the Lane County Circuit Court, Windsor Calkins, son of S. M. Calkins, was appointed Board counsel—the same year the Eugene Water Board was redesignated the Eugene Water & Electric Board. Windsor Calkins has served in the position of legal counsel to the Board since 1949, with his son, Win, now sharing the responsibility of providing legal guidance to the municipal-utility operation.

Windsor Calkins, his father, S. M., and his son, Win, are not the only members of this family who have served the community. Windsor's uncle, W. W. Calkins served as a member of the City Council's Fire and Water Committee before the formation of the Water Board and was active in that process.

W. W. Calkins is listed in the 1910 Polk city directory as Winsor Calkins, partner in the law firm of Calkins & Calkins. The other principal of that firm was Sylvenus M. Calkins, more commonly known as S. M. Calkins. The present-day Windsor Calkins spelled his own name "Winsor" rather than "Windsor" until the time of his entry into military service in World War II. It was then that he discovered that his birth certificate identified his given name as "Windsor."

An attempt was made to thwart the appointment of Windsor Calkins in 1949 when Board member Stan Darling, a Eugene attorney, made an effort to have the Board appoint the firm of Bailey & Hoffman as legal counsel. Failing for lack of a second to his motion, Darling then sought Board appointment of Philip Lowry to the post. Lowry was a member of Darling's law firm. Darling's motion to appoint Lowry was seconded by Board member and University of Oregon Professor Howard Taylor. However, the motion was defeated with Board members Will Robertson, Lionel Trommlitz and Paul Green voting against Darling and Taylor. The appointment of Calkins then went forward.

(Left) G. F. Skipworth, earliest attorney to provide legal services to the Board while serving as city attorney; later Lane County Circuit Court judge for whom juvenile detention center is named (Courtesy Register-Guard). (Right) S. M. Calkins, city attorney and attorney to the Board 1925 to 1946 (Courtesy Windsor Calkins).

BORROWING AUTHORITY ACHIEVED

William East, whose tenure as Board counsel was relatively brief, was instrumental in guiding the Board toward obtaining voter approval for borrowing up to $500,000 from local banks on a short-term basis without the requirement of an election. This authority was expanded in 1976 when a charter amendment permitted the Board to borrow an amount up to five percent of retained earnings. The special borrowing authority has never been used, but its availability has provided comfort to the utility management which is constantly faced with the possibility of costly emergencies. After serving as counsel to the Board, East had a distinguished career as a jurist both in Lane County and on the federal bench.

After Calkins was appointed Board counsel, General Superintendent-Secretary Ray Boals asked him if he planned to attend all Board meetings. Calkins replied that he would not, explaining that the Board's business was that of policy making and that the presence of legal counsel would hinder the policy-making process. Rather, he told Boals, legal counsel should be knowledgeable in the affairs of the utility operation and be available on short notice to provide legal advice. A 37 year relationship has demonstrated the wisdom of Calkins' position.

As a true professional, Windsor Calkins knew when it was prudent to call upon legal specialists to provide assistance to the Board. In 1964 the Portland-based law firm, Schwabe, Williamson, Wyatt, Moore & Roberts (today's firm name) was engaged as special counsel to participate with Calkins in the defense of the Eugene Water & Electric Board in a lawsuit filed in Federal District Court by Peter Kiewit & Sons over claims arising out of the Carmen-Smith construction project.

STAGE SET FOR TROJAN PARTICIPATION

When EWEB became interested in thermal-nuclear generation in the mid-1960s, John Schwabe, a principal of the Portland firm, was asked to seek state-level legislation which would make it possible for Oregon public entities to jointly participate with private companies in the construction of large thermal projects. This was accomplished and the stage was set for later participation by EWEB with Portland General Electric Company and Pacific Power & Light Company in the Trojan Nuclear Project as part of Phase I of the Pacific Northwest Hydrothermal Program. The firm negotiated the necessary contracts with the two Portland-based private utilities and the Bonneville Power Administration. This process involved obtaining Congressional approval and Internal Revenue Service rulings necessary to clarify that bonds to be sold by EWEB would be tax-exempt. The final contractual arrangements were completed in October, 1970 and therafter, EWEB sold several series of bonds to finance its share of the Trojan Project.

Following completion of the contractual arrangements for Trojan, the firm assisted EWEB in regional negotiations for Phase II of the Hydrothermal Program. However, when BPA, because of several environmental lawsuits, was prevented from participating in the program, the region sought federal legislation which would expand BPA's ability to acquire resources to meet the needs of its customers. EWEB and the Portland law firm were active in the formulation of this legislation but ultimately opposed it as being unworkable and infringing on local control of utility matters. The legislation was signed into law in December, 1980.

In the wake of negative voter reaction to nuclear generation in 1970 EWEB became a leader in pursuing conservation and renewable resources. During 1975 the firm assisted Herbert Hunt, EWEB power resources manager and John Brown, EWEB treasurer, in negotiations with Weyerhaeuser Company for the installation of a cogeneration plant in Springfield. This resulted in construction of the 51 megawatt plant and as part of the arrangements, EWEB entered into contracts with Pasadena, Glendale and Burbank to sell the power until it was needed in Eugene.

(Left) William G. East, first attorney to represent the Board independent of the City Council (Courtesy Register-Guard). (Right) Windsor Calkins, appointed Board counsel in 1949 (Kennell-Ellis photo).

Early in 1980 the firm assisted General Manager Keith Parks in negotiating new power sales contracts with BPA. The firm was also active in BPA rate proceedings which, as a result of passage of the Regional Act, took on an entirely new meaning for the region and the public entities. Along with other major generating utilities, EWEB was active in closely observing the BPA rate-making process.

As EWEB became a leader in conservation, the firm assisted EWEB in drafting conservation contracts with BPA. In 1983 and 1984 EWEB negotiated a unique arrangement with BPA whereby EWEB would finance conservation through the sale of tax-exempt conservation bonds with the resulting conservation savings credited to BPA under the Regional Act. Both the Portland and Washington, D.C. offices of the firm worked with the Senate finance staff to maintain the statutory authority of BPA to provide financial backup for the EWEB conservation bonds.

(Left) Win Calkins, shares responsibility of providing legal guidance to Board (Kennell-Ellis photo). (Right) Rockne Gill, member of Portland firm of Schwabe, Williamson, Wyatt, Moore & Roberts, special counsel to the Board.

For a number of years Rockne Gill of the Portland law firm has taken a principal responsibility in the firm's relationship with EWEB. Work assignments have been divided among Gill, Jay Waldron, John Crawford, Larry Cable, Jim Benedict and Roy Lambert.

Windsor Calkins and the Board looked to New York City as bond financing became increasingly complex. As the 1961 Carmen-Smith Series A bonds were being considered, the firm of Wood, King, Dawson & Logan was engaged as bond counsel. This established a relationship which would endure for many years. John Dawson, active in the early years of the relationship, was capable of explaining complexities of the bond market in straightforward, understandable terms—a tradition which has been continued by other principals of the firm. Others who provided valuable assistance to the Board were Leroy Love and Brenden O'Brien. In recent times Edward J. McCormick assisted with work on the 1982 water system bonds and the conservation bonds.

Other attorneys have been engaged for special assignment. In recent times Akin Blitz has served the Board well in certain cases relating to employment matters.

EWEB staff members, when visiting Windsor Calkins' office for legal guidance, are treated to a rare combination of legal wisdom, an understanding of local history—particularly that of EWEB, common sense and a fine wit. Win Calkins over the past several years has demonstrated the fine qualities possessed by his father. Both are true gentlemen.

The Board's legal counsel has provided continuity which has spanned several administrations. The value of this contribution is beyond measure.

CHAPTER NINE
The Board

DURING THE HEIGHT OF THE nuclear campaign in 1970, a proposal was made to members of the Eugene Water & Electric Board that extreme measures be taken in the Board's fight against the Eugene Future Power Committee which opposed nuclear power and EWEB's plans to build a plant. Suggested measures would include slick, Madison Avenue techniques.

The Board members declined. President John Tiffany and Vice President Kenneth Kohnen, speaking for the entire Board, stated that during the remainder of the campaign, they would continue to tell the public only the facts and truth as they knew them and would not resort to tactics that might mislead voters.

The opponents of nuclear power had a decided edge. The *Eugene Register-Guard* had employed an outstanding young writer, Gene Bryerton, whose initial, singular purpose was to help thwart EWEB's plans for nuclear-plant construction. Local media gave favorable, often redundant, coverage to several antinuclear speakers of national prominence who came to Eugene.

There was basis for bitterness in the Eugene business community. But members of the Board would not be swayed from their course of truth and decency. Their refusal exemplified the high level of integrity which historically has characterized the Board.

RAPID CHANGE IN INDUSTRY

President Tiffany not only deported himself in exemplary fashion during the nuclear campaign. He would successfully meet challenges of great importance as events of the 1970s unfolded. The nature of Eugene's utility operation and the utility industry in general was changing rapidly. The Board, which historically had been dominated by representation from the business community, began to shift dramatically in composition. Electric energy could no longer be considered an ever-abundant commodity.

John Tiffany provided the leadership that would guide the utility operation to prominence as a pioneer in energy conservation and the investigation of alternative, renewal energy resources. He ably conducted Board meetings which, in the early 1970s, were often characterized by sharp differences of opinion. In his quiet, businesslike approach to increasingly complex issues, Tiffany strengthened EWEB's position in the community and the region.

During his years with the Board, Tiffany did much more than attend and conduct meetings. He did his homework well and became keenly knowledgeable in complicated utility matters.

For many years Tiffany operated a steam-driven pleasure boat on the Siuslaw River. He enjoyed taking friends for rides along the river, showing them the operation of the oil-fired boiler that powered his boat. This love for machinery could be observed in Tiffany's deep interest in utility technology—particularly with respect to new concepts.

John A. Tiffany's service with the Eugene Water & Electric Board spanned 22½ years including three years as vice president and 17 as president. This is the longest service rendered by any Board member in the organization's history. Tiffany retired from the Board at the end of December, 1984. His intelligent, ever-ebullient wife, Pauline, has had a distinguished career in business education and management. She has served as regional president and as a national officer for the American Society for Personnel Administration.

BOARD REACTS TO COUNCIL MOVE

It was August 9, 1976. The Eugene City Council had under consideration a revision of the City Charter which was to be submitted to the voters. Save for the original Water Board, Board members historically had been elected by the people with vacancies which occurred being filled by the Board itself. At an earlier meeting of the City Council, in an informal action, Councilors Neil Murray, Wickes Beal, Ray Bradley and Eric Hawes voted to eliminate the existing practice in which the Board acted on its own to select replacements when vacancies occurred. Councilors Gus Keller, Tom Williams and Edna Shirey had voted at the earlier meeting to retain the existing method.

Reaction from EWEB was swift and emphatic. Board members John Tiffany, Taylor Ramsey, John Reynolds, Richard Freeman and Calvin Schmidt were in attendance at the August 9 Council meeting as were many other EWEB supporters. Taking the lead as spokesman for the EWEB side was University of Oregon Law School Dean Emeritus Orlando John Hollis who addressed the Council with eloquence. Fifteen

John A. Tiffany; service on Board spanned 22½ years. (EWEB files)

people, who had signed up to speak for EWEB, yielded their three-minute time allotments to Hollis, providing him ample time to offer his comments. Dean Hollis praised EWEB for its contribution to the community and recalled that Eugene Mayor Francis Wilkins, in 1906, proposed a Water Board which would "operate independently beyond all political strife."

"Why is there any thought or disposition to depart from a policy that has been followed with the greatest success in this city for 65 years?" Hollis asked the Council. "How many cities could you name that could put down this kind of success story? Why meddle with it?"

In the burst of applause that followed, Mayor Les Anderson rapped his gavel for silence, but did not seem displeased. Later that evening, the Council voted unanimously to leave the existing practice of replacing Board members intact.

Hollis knew of which he spoke. He had served on the Eugene Water Board from 1931 to 1941 and had remained a good friend of the utility operation since his time of service.

Dean Hollis would serve the Board in yet another important capacity. When the Board held public hearings in the early 1980s required by the National Energy Act, it looked to Hollis to serve as presiding officer. He carried out this responsiblity in an exemplary manner. In a lighter aspect of his relationship to the Board, Hollis served as master of ceremonies at the retirement parties for General Managers Byron Price and Keith Parks as well as that of Board member and president John Tiffany.

The influence Orlando John Hollis has exerted goes vastly beyond the tutelage rendered to thousands of practicing attorneys and countless jurists. Just as he was an example to his faculty and students, he continues to be an example of integrity and dignity in the community he loves. In a time when education in the United States is being critically examined, Hollis represents a tradition of academic excellence. Without question, he has had great impact upon higher education generally.

Dean Orlando John Hollis is the son of the late Moody M. and Amelia Hollis. Moody M. Hollis was one of 15 children born to John W. and Barbara Hollis. He worked for many years in Eugene as chief clerk and then assistant agent for Southern Pacific Railroad.

When John Tiffany decided to retire from the Board and step down as its president, he could be confident that high-quality leadership would continue. That confidence rested in the presence on the Board of Camilla P. Pratt, a veteran with Board experience dating back to January, 1977. She was elected president in January, 1985.

Camilla Pratt, in January, 1979, had declined nomination for the Board presidency. In an effort based upon what were perceived at the time as honest philosophical differences, newly seated Commissioners John R. Bartels and Jack J. Craig nominated Camilla Pratt for the top Board post.

Declining the nomination, Pratt added, "I feel the uncertainties and problems that our community and the region are facing with regard to electric energy require a steady, experienced individual for president of

Dean Orlando John Hollis. (Kennell-Ellis)

the Board of Commissioners, a fair-minded person who has the confidence and respect of the community."

Following her statement, Pratt nominated John Tiffany for reelection as president. By a vote of three to two, Tiffany would continue to provide leadership to the Board.

Notwithstanding philosphical differences that might have existed in January, 1979, as important issues of the 1980s emerged, there was, more often than not, unanimity in positions taken by the Board.

When Camilla Pratt was seated on the Board in 1977, she probably was the best prepared candidate in the Board's history. She had attended countless Board meetings and had become knowledgeable in the problems and affairs of the utility operation.

Pratt has maintained a special relationship with the staff and employees. Her frequent and sincere expressions of appreciation have provided encouragement and a sense of pride. Her attendance at employee awards banquets and the many occasions on which she has visited hospitalized employees demonstrate her concern for the human side of the municipal utility.

CONCERN FOR CUSTOMERS

This same concern can be observed in Camilla Pratt's sensitivity toward problems and questions expressed by customers. In Board dialogue she makes frequent reference to issues that have been raised by her constituents. She gives her undivided attention during public input at Board meetings.

Possessing a high level of intellectual integrity, Pratt does not hesitate to revise her position if her research so dictates. On October 9, 1978, she told other Board members that she had reversed her position "180 degrees" on lifeline rates.* An early proponent of such rates, she changed her position only after following the results of implementing lifeline rates in California and elsewhere. Experience with the rate structure, she found, demonstrated that lifeline rates did not promote equity nor more efficient energy use; in fact, these rates disproportionately shifted the rate burden to commercial and industrial customers.

As the struggle among levels of government, publics and institutions intensifies, it is the task of the Eugene Water & Electric Board not only to survive, but to survive and maintain its traditional standards of excellence of service to its customers. Camilla Pratt is well-suited to provide leadership in meeting this challenge.

Camilla Pratt and the community can look to the other Board members for able assistance in the governing of the utility operation. Dennis L. Solin was elected vice-president in January, 1985. He has served on the Board since 1982 and is a certified public accountant.

Dr. Sarah S. Hendrickson, the first physician to serve on the Board, has often demonstrated sound judgement. She has served since 1981.

Jack J. Craig, with Board service since 1979, has had long experience in governmental and political affairs. He has served the community well in various capacities as a journalist.

*"Lifeline rates"—a concept in which a first block of electricity—an amount deemed necessary for basic needs—is provided at low cost. A major problem with this concept is that it is contrary to rate design based on cost of service.

James E. Reinmuth was seated on the Board in January, 1985. Dean of the School of Business Administration at the University of Oregon, Reinmuth is the sixth university faculty member to serve on the Board, being preceded by Orlando John Hollis, Howard Taylor, Egbert Wengert, J. Orville Lindstrom and John Reynolds.

Camilla P. Pratt. (Dann's Photo)

What of the early Board members? The original five were named by city ordinance. Probably they were selected because of their prominence as businessmen in the thriving community of Eugene.

Robert M. Day, who was elected temporary chairman and first president, was co-proprietor of Day & Henderson, an undertaking and furniture establishment—a common combination in those times.

Martin Svarverud, a native of Norway, was elected vice president of the first Board. He was senior partner and manager of the Eugene Real Estate and Investment Company. Previously, he had operated a hardware and implement business.

Ernest U. Lee, a banker, would later serve as county clerk and as mayor of Eugene in 1925-26. Burwell B. McKinney, who would remain on the Board until 1916, was in real estate.

Clement S. Frank had come to Eugene in 1888 and had become a successful merchant. Both Frank and Lee served on the Board until 1919.

As mentioned in earlier pages, Joseph W. McArthur defeated Martin Svarverud in 1913 and following his service as a Board member which spanned 10 years, would later become general superintendent-secretary.

J. S. Drew (1917-1922) and E. M. Drew (1923-1928) were of the same family and were in the construction business. Elmer K. Wheeler (1916-1922) was vice president of Eugene Savings & Loan.

Alex T. Cockerline (1919-1923) was with Cockerline & Weatherbee, a dry-goods business. Harry O. Bowen (1919-1920) was manager of Allen & Lewis, wholesale grocers. Lewis D. Pierce (1921-1922) operated a delicatessen called "The Table Top."

Percy W. Brown is the earliest of the 28 out of 49 Board members known personally to the writer. He served from 1923 to 1940. A certified public accountant, he is remembered as a man of uncommon dignity. He served for a period as Board president during his service on the governing body.

Walter P. Fell served from 1922 to 1934. He is remembered by Dean Orlando Hollis as a "fine, even-tempered gentleman." Fell preceded Percy Brown as Board president and while on the Board, was manager of Mason Ehrman & Company, wholesale grocers.

Bruce B. Brundage (1923-1925) was cashier of the Bank of Commerce. Marion Veatch (1925-1932) was an undertaker and Vern D. Scobert (1929-1931) was in the building material business.

Orlando John Hollis, who served from 1931 to 1941, has been discussed previously. Cyrus J. Fulton (1933-1936) was from time to time involved in men's clothing stores. He also was well-known on the Pacific Coast as an artist. He had been raised in The Dalles area and had become well-acquainted with some of the Columbia River Indian tribes. His paintings of Indians and Indian scenes drew broad attention.

William R. "Will" Robertson served continuously on the Board from 1934 to 1954. During this period he was elected three times and ap-

(Left) Dr. Sarah S. Hendrickson; (right) Jack J. Craig. (EWEB files)

(Left) Dennis L. Solin; (right) James E. Reinmuth. (EWEB files)

pointed by the Board twice. Warmly remembered by those who knew him, he served as Board president and made a lasting contribution to the utility operation. In 1949, while he was president, he prevailed in his support of the appointment of Windsor Calkins as Board counsel. Robertson was in the insurance business.

Alexander L. Williamson (1936-1944) was in the plumbing, heating and ventilating engineering business and Albert J. Gillette (1941-1943) was local agent for Southern Pacific Railroad.

Paul D. Green (1941-1951) continued to maintain a close attachment to EWEB after his time of service. He was a men's clothing store owner.

Lionel W. Trommlitz, or "Tromm" as he was called by friends and associates, served from 1944 to 1964. In his capacity as president, he provided leadership as the Board undertook the biggest construction job in its history, the Carmen-Smith project. Trommlitz spent many years in radio station management.

Howard R. Taylor (1945-1954) was a psychology professor at the University of Oregon. Stanley R. Darling (1949-1954) was a practicing attorney.

A banker, Oluf A. Houglum served from 1951 to 1962. A fine gentleman, Houglum guided the Board toward selling revenue bonds for the Carmen-Smith project as opposed to more traditional general-obligation bond financing.

Calvin R. Smith retired as J. C. Penney store manager while serving on the Board. He, along with Trommlitz, gave leadership during the planning and construction phases of Carmen-Smith. His time of service was from 1954 to 1967. After having been appointed in February, 1954, he successfully stood for election three times.

The first Board, March 11, 1911; (left to right) Clement S. Frank, B. B. McKinney, Robert M. Day, Ernest U. Lee, Martin Svarverud. (EWEB files)

(Left) Alex T. Cockerline (1919-1923); (right) Lewis D. Pierce (1921-1922). (EWEB files)

(Left) Percy W. Brown (1923-1940); (right) Walter P. Fell (1922-1934). (EWEB files)

(Left) Marion Veatch (1925-1932); (right) Vern D. Scobert (1929-1931). (EWEB files)

(Left) Cyrus J. Fulton (1933-1936); (right) William R. "Will" Robertson (1934-1954) (EWEB files)

LEGEND IN HIS OWN TIME

The venerable Earl L. McNutt served from 1954 to 1965. A man of the earth, of the mountains and trees, and a life hard worked, McNutt was a legend in his own time. His was a simple, but straightforward and forceful wisdom. On several occasions during Board meetings, a stranger might have thought that McNutt had nodded-off—such deference, after all, being accorded Justice Holmes and others of advanced years. But when it was time for a motion, it was often McNutt who made it and it was made from information and a full understanding of the issue at hand. McNutt's multifarious career included construction, farming and property ownership.

Egbert S. Wengert (1955-1956) served a short time on the Board. However, he was keenly interested in the Board's problems and issues and spoke of them often. In the early 1950s, Wengert, as head of the political science department at the University of Oregon, undertook the first formal executive-training program for the Oregon state government. On one occasion, as he was conducting an evening session in the basement of the Oregon State Library, he suggested to his group that it was sometimes difficult to distinguish between engineering and political decisions. This immediately raised the ire of Deputy State Highway Engineer Forrest Cooper. Reinforcing his contention, Wengert then cited a highway in the Midwest which had been built with 10-mile, al-

(Left) Alexander L. Williamson (1936-1944); (right) Albert J. Gillette (1941-1943). (EWEB files)

(Left) Paul D. Green (1941-1951); (right) Lionel W. "Tromm" Trommlitz (1944-1964). (EWEB files)

ternating segments of macadam and concrete, then asking the group if this could have been an engineering decision. This was too much for the fiery Cooper. However, the exchange was short and both men seemed to respect one another as a result of the episode. Certainly both held the respect of their colleagues. Cooper, a man of high reputation, would become State Highway Engineer.

Julio Silva (1956-1959) possessed a down-to-earth friendliness that impressed many of the employees. Following daytime Board meetings, he would often remain in the building, sitting and talking to employees of various levels. Silva was a strength on the Board during the Carmen-Smith planning phase. For several years, he owned the Chevrolet dealership in Eugene.

J. Orville Lindstrom served from 1959 to 1964. Business manager at the University of Oregon, he had a keen understanding of fiscal affairs and made a valuable contribution to EWEB. When the Board was considering long-term disability insurance in 1964, Lindstrom raised the issue of protecting the retirement benefits of disabled employees—a matter which had not been considered at all in the staff proposal. At his insistence, the proposal was revised and brought back to the Board which approved it. Thus the some 30 EWEB employees who have become disabled have received a benefit of critical importance to them.

John A. Tiffany is discussed in earlier pages. Kenneth H. Kohnen

(Left) Howard R. Taylor (1945-1954); (right) Stanley R. Darling (1949-1954). (EWEB files)

(Left) Oluf A. Houglum (1951-1962); (right) Calvin R. Smith (1954-1967). (EWEB files)

served from 1964 to 1970. A certified public accountant, he made an important contribution to EWEB.

Trygve S. Vik was a member from 1964 to 1972. He is a successful building contractor. Ehrman V. Giustina, a Eugene lumberman, served from 1965 to 1972. Richard F. Olson, long involved in concrete products manufacturing, served from 1968 to 1973.

EWEB'S MOST DIFFICULT TIME

Kohnen, Vik, Giustina and Olson, along with President John Tiffany, formed a team that saw EWEB through its most difficult time—the nuclear issue which reached its height in 1970. In the mid-1960s nuclear power appeared to be a viable, safe energy source. The Board was well-intentioned as it went forward with a thermal generation bond election in 1968. It was the consensus of Northwest utilities and the Bonneville Power Administration that several new thermal plants would be needed in the region.

As the Board began to study various sites for a plant location and the technology involved, local opposition to a Eugene plant developed. This opposition, which was spearheaded by the Eugene Future Power Committee, was successful in seeking and achieving a four-year moratorium on construction of an EWEB-sponsored nuclear plant. At the election held on May 26, 1970, the moritorium was approved by the voters.

It would later be determined that forecasts of power needs that had been developed by the utilities and BPA were unrealistically high. Further, all of the plants outlined in the utilities' hydro-thermal program would not be needed—certainly not before the turn of the century or beyond.

But EWEB commissioners in 1970 were acting in good faith. To their credit, once the moratorium was fact, they quickly regrouped and prepared to meet other challenges that would emerge in the 1970s. Kohnen, Vik, Giustina, Olson and Tiffany were men of dignity who were highly dedicated to the public trust placed in them.

Richard F. Freeman, seated on the Board in January, 1971, was both a businessman and a University of Oregon research associate. He was not bound by any persuasion other than careful analysis of the issue at hand. Early on, he expressed his concerns about polychlorinated biphenyls (PCBs), a suspected carcinogen that had been added to coolant oil for electrical equipment by utilities throughout the nation. His homework went well beyond the heavy-reading requirements which burdened commissioners. Richard Freeman's service on the Board was from 1971 to 1980.

Gerald S. Moshofsky served one year in 1972 after having been appointed to fill the unexpired term of Trygve S. Vik. He was in the wood-components and wholesale lumber business. His business interests later took him to Iran and following the 1979 hostage crisis in that coun-

(Left) Earl L. McNutt (1954-1965); (right) Julio Silva (1956-1959). (EWEB files)

(Left) J. Orville Lindstrom (1959-1964); (right) Kenneth H. Kohnen (1964-1970). (EWEB files)

(Left) Trygve S. Vik (1964-1972); (right) Ehrman V. Giustina (1965-1972). (EWEB files)

(Left) Richard F. Olson (1968-1973); (right) Richard F. Freeman (1971-1980). (EWEB files)

try, he provided Eugene service clubs with a cogent and informative account of the events leading to the crisis.

Taylor Ramsey served from 1972 to 1978. Gifted with an uncommon sense of humor, Ramsey combined this attribute with an early-on, deep understanding of utility issues. During his time of service, Ramsey was president of United Pipe and Supply Company and principal in Ramsey-Waite Company. In recent times he has served as president of the Eugene Chamber of Commerce. He also served two years as chairman of that body's energy committee.

An active member of the Eugene Future Power Committee which helped defeat EWEB's proposed nuclear plant, John S. Reynolds was elected to the Board in November, 1972 and served from 1973 to 1976. Keenly intelligent, Reynolds was quick to grasp utility issues. Reynolds is a professor of architecture and director of the Solar Energy Center at the University of Oregon. Since his time of service, he has maintained his association with EWEB by using the utility operation as a subject of study for his students.

Calvin L. Schmidt, a building contractor, served from 1973 to 1978. He, along with Commissioner Taylor Ramsey, constituted a special committee which had as its purpose the study of EWEB operations. The two commissioners found operations to be in good order and made relatively few recommendations. Schmidt, as was the case with Ramsey and Reynolds, developed a thorough understanding of utility matters.

Camilla Pratt has been discussed in earlier pages. Thomas L. Teutsch, as was the case with Gerald Moshofsky, served only one year (1978). Even with such short tenure, Teutsch was beginning to understand Board issues and had he remained on the Board, he would have made an important contribution. He is in the financial-planning business.

John R. Bartels, who served from 1979 to 1982, is a journalist. Of anti-nuclear persuasion, he proved to be level- headed and articulate with regard to general-utility issues. Joining with Commissioner Craig in an effort to unseat President Tiffany in 1979, Bartels, as he announced his resignation on October 4, 1982, praised Tiffany's leadership.

SUCCESS CONTINUES TODAY

Are there attributes that have prevailed through 75 years of Board history? There appear to be. The Eugene Water Board was created because of a critical need in the community. The typhoid epidemic of 1906 was a factor. But in a broader sense, the community seemed to understand that its future would require, if not depend upon, stable, business-like delivery of utility services. Municipal ownership, with a separate governing body, was deemed in those times to be the best means to achieve this end. The concept was successful then and that success continues today.

(Left) Gerald S. Moshofsky (1972); (right) Taylor Ramsey (1972-1978). (EWEB files)

(Left) John S. Reynolds (1973-1976); (right) Calvin L. Schmidt (1973-1978). (EWEB files)

(Left) Thomas L. Teutsch (1978); (right) John R. Bartels (1979-1982). (EWEB files)

The Board, from its very first meetings to the present day, has had as its foremost purpose the providing of good, reasonably priced service to its customers. It has historically carried out this charge with integrity and dedication. The technology is vastly more complex as is the political and governmental environment, but the purpose remains essentially unchanged.

Eugene's municipal utility system is the envy of communities throughout the region. More importantly, it is an organization of which its own community can be proud.

CHAPTER TEN
Some Employees

IT HAS BEEN SOME YEARS NOW. Norman Johnson, former water operations superintendent, visited EWEB headquarters. His purpose, in addition to seeing old friends, was to show EWEBers his motor home. A converted Trailways bus, the vehicle was a marvel of craftsmanship.

It was the writer's privilege to accompany Chief Draftsman Ron Davis on a tour of the motor home and to introduce him to Johnson.

Each of the two men noted immediately that the other had lost an arm. After briefly discussing the time and circumstance of loss, Johnson, unhesitatingly, said, "the handicap is all in the mind isn't it?"

"That's right," Davis answered with his characteristic smile.

If we're fortunate, we encounter such eloquence perhaps two or three times during our lives—eloquence born not of arts and letters, but of life.

Perhaps the quality, or character, described above is what Board President Camilla Pratt had in mind when she recently referred to EWEB employees as "highly principled."

If there is a single trepidation that troubles the writer of this kind of volume it is that of seeming to ignore some employees who have made important contributions. In foregoing pages, when employees have been part of a particular activity, they have been mentioned. Otherwise, emphasis has been placed on decision-making and management activities of the Board and its chief executive officer.

Nevertheless, there are individuals other than Board members and top managers who have contributed a great deal to the work of the Board. Some are mentioned in this chapter.

Byron K. Taylor, including time spent in the military during World War II, was in the Board's service for nearly 40 years. A civil engineer he was involved in much of the construction which took place during the McClain administration. By the time of the Boals era, Taylor was a strong member of the utility operation's management team.

WORTH NOT MEASURED BY CREDENTIAL

Taylor measured the worth of an employee not by credential. He respected hard work and craftsmanship. It was important to him that those working along the McKenzie have an understanding of the river along with its forces and vagaries.

When utility management salaries began to assume magnitude during the early 1960s, Taylor remarked that it required a lot of utility bills to pay his own salary. He was deeply sensitive about this with commensurate sensitivity to the matter of providing good service to customers.

The quality of Taylor, the manager of operations, could be seen in his operating supervisors. The writer and Taylor discussed at length the matter of the Hayden Bridge Filtration Plant and the fact that with Ben Barnts in charge, neither the Board nor the general manager need be concerned about the quality of water, day or night, that flowed through the lines to Eugene—a tradition of dependability that has been carried forward by Floyd Farstad and Doug Wise. Similarly, the delivery of water to meters within the Eugene system could be relied upon—due in large part to a tradition of service represented by Bevery "Bud" Barnts, brother of Ben and lead water serviceman.

SYSTEM RELIABILITY HIGH PRIORITY

This same dependability and caring about customers could be seen throughout Taylor's department. Chief Engineers Ralph Martin and Charles Strong, engineers Dean Hodges, Walt Ruppel, Jim Howell, Don Robinson, Francis Whitney and Dean Axtell, electric distribution engineers Mel Damewood and Bill Fulps, Chief Surveyor Dan McCornack and Chief Draftsmen Art Bevers and Dean Wentworth collectively viewed system reliability as a high priority. Untold sums of ratepayer money were saved by the careful management of power resources carried on by Miller Evans, Hal Worcester, John Fuller and Barbara Lamb with this work being continued today by Vaughn Scales and Roger Nelson.

During construction of the downtown underground electrical network in the early 1960s, "Mule Gang" Foreman Glenn Haney saw to it that business proprietors along affected streets were properly informed in advance as to when streets would be opened. Construction was often scheduled on weekends in order to hold public inconvenience to a minimum.

To those employees not well acquainted with Byron K. Taylor, he may have seemed distant at times. But those who knew him also knew his deep feeling of dedication to the work force, customers and the utility operation. A fine gentleman in every regard, he retired in May, 1969.

Grace Shaffner joined the Eugene Water Board in July, 1926 and retired in January, 1968. She worked in many capacities in the accounting department and in later years, was involved in the billing operation.

Shaffner witnessed much of the dynamic construction period of the McClain years. She was deeply involved in the post-World War II period of growth and was a valuable employee as accounting functions were placed on computer in the mid-1960s. She, along with such colleagues as Nila Millsap and Marilyn Ballard, was a strength in the utility.

Walter J. Moore began working for the Eugene Water Board as a part-time employee in 1919, having graduated in engineering at the University of Oregon in 1909. In 1925 he was appointed full-time civil engineer and became water operations superintendent in 1932. He retired in 1952.

EWEB photograph files contain many pictures of construction projects taken by Moore. He was involved in design and engineering work on several water and electric projects including Skinners Butte Reservoir, steam plant, Hayden Bridge-Eugene water line and the Leaburg hydroelectric project.

Lucille Norton joined the organization in February, 1928. Working principally in a secretarial capacity, she became knowledgeable in Board policies and records.

Norton, in her later years with the Board, also became known as the house grammarian. She not only knew Board policy history, but also knew how to set up and carry out complex procedures implementing policy.

Post World War II; (seated left to right) Hugh Currin, Ford Northrop, Ray Boals, Walter Moore; (standing left to right) Ralph Patterson, Jim Howell, Ralph Martin, Don Robinson, Vern Poindexter, Norman Johnson. (EWEB files)

Following over 42 years of service, Lucille Norton retired in 1970.

Carrying on the tradition of excellence in secretarial work was Phyllis Wald who was employed from 1962 to 1981. First working in administration, she later spent many years as secretary to the treasurer and the finance and accounting division.

George Easton and Hugh Currin joined the Eugene Water Board in 1923 and both would have distinguished careers. Easton devoted his years to the hydro systems at Walterville and Leaburg. By Byron Taylor's definition, Easton understood the lower reaches of the river.

Hugh Currin is the only person for whom a facility of the Eugene utility is named—the Hugh P. Currin Substation, located along Interstate 105 east of Coburg Road. The use of the designation "Matlock Station" for the Walterville generating plant was discontinued in the 1920s and the Carl A. McClain Filtration Plant was taken out of service in the early 1950s.

Currin, a graduate in electrical engineering at the University of Oregon, is credited, along with Vernon J. Poindexter, with the design and construction of the electric system. A pragmatist with a sense of humor, Currin in later years often referred to the accumulation of useless items common to most households. "Have a fire every few years," he would admonish.

Byron Taylor (center) congratulating Leland Robe at the latter's retirement party in 1968; Fred Stuhr at left. (EWEB files, H. Gross photo)

"MAYOR OF LEABURG"

George Easton, possessing limited formal education, became well-known in his community for his business acumen—particularly in fiscal and property matters. Known for a period of years as unofficial "mayor" of Leaburg, Easton also served as a member of the Springfield School

Retirement party in 1956 for Ralph Russell (center); Russell was first employed by the Eugene Water Board in 1912; Darwin Spicer on left, Ray Boals at right. (EWEB files)

District 19 board from 1953 to 1957 and as a member of the district's budget committee before that time. While associated with the district, Easton played an important role in the acquisition of 12 school-building sites. Checking his diary recently, he found that the Walterville penstock rupture (photo in Chapter 3) occurred at 5:30 p.m., August 11, 1923—30 minutes after men working on the new plant left work for the day.

Hugh Currin retired in 1954 and George Easton in 1962.

(Standing left to right) Fred Gadsden, Vern Poindexter, Miller Evans; (seated) Ida Peil, cook at EWEB cafeteria. Occasion is retirement party for these employees in 1968. (EWEB files, H. Gross photo)

STRENGTH OF THE ELECTRIC SYSTEM

Vernon J. Poindexter began his career in November, 1928 and retired in June 1968. For many years he was the strength of the electric system and few, if any, projects in the electric department were undertaken during the 1930s, 40s, 50s and 60s which did not benefit from his involvement.

1969; (left to right) Clarence Hill, hydro systems supervisor; Harold Graham, chief dispatcher; Harold Worcester, power resources; Leonard Boyland, overhead lines supervisor. (EWEB files)

Poindexter, like Byron Taylor, sometimes seemed distant to his employees. But as was the case with Taylor, Poindexter had a deep regard for his work force. Often, his solution to a difficult administrative or personnel problem was more balanced and humane than that arrived at by those of us who thought we were "experts."

At the time of the Columbus Day storm in 1962, and during other periods of storm damage, outage times in the EWEB system were far less than those of some neighboring utilities. This was attributable to the high level of electric system integrity which was, in large part, the result of Poindexter's efforts.

Notable in the employee leadership in the attempt at union organization in the mid-1950s were Richard D. Gall and Lynn H. Baxter, both employees in electric operations. Recrimination against such "radical" behavior might have occurred, but didn't. Rather, the organization nurtured the talents of Gall and Baxter and as a result of this process, Gall replaced Poindexter as electrical operations superintendent and served very capably until his retirement in 1984 (the position was occupied for a period by Howard Stenshoel on a temporary basis following Poindexter's retirement). Had Baxter's career not been cut short by his early, untimely death, he, too, would likely have entered management ranks.

During Poindexter's time as superintendent, he and the electric department were assisted by many key personnel. John Scofield served for several years as assistant electric operations superintendent. Oscar A. Adams supervised the electric meter shop, assisted in turn by Ross McClanahan and Albert Kronberger. Glenn Haney, mentioned earlier, was assisted in supervision of the underground electric and substation maintenance activity by Loyd Coons, Tom Davis and Ivan "Ink" Smith. Fred Gadsden and Harold Graham served as chief power dispatcher and both Ralph Russell and Darwin Spicer served as steam plant supervisor with this activity carried forth by Clark Lynch and Everett Jordan. M. B. "Tiny" Eaton, Neil "Pappy" Sleeper and Walter Larwood supervised overhead lines construction and maintenance with Howard Stenshoel and Art Delmore eventually joining this activity. Electric servicemen included Thelmer Nelson, Willis Dotson, Harold "Ike" Lusk and Louis Polley. Dotson, Lusk and Polley would become supervisors. Notable among crew foremen were Leonard Boylan, John Hauck and Al Guy, all of whom would become supervisors. Other crew foremen were Laudie Kralicek and Sam Pakulak.

1969; (left to right) Curt Sprecher, Harold King, Ray Barlow—all of the accounting department. (EWEB files)

Poindexter's final assignment at EWEB was to serve as special consultant to General Manager Byron Price. In this capacity he conducted a survey of the overhead-lines system.

Kenneth W. Rinard joined the organization in April, 1949. During his many years as water operations superintendent, he was ably assisted by the assistant superintendent, Winston Berkeley, Ben Barnts, filtration plant supervisor, Lawrence Lauderdale, office coordinator, and Harold Buell, crew foreman. Other key field personnel included Ted Wayne, Hobart Campbell, Earl Buell, Murray "Bud" Nadeau, Jack Davis, Tom Ritchey and Carl "Oly" Johnson, father of recently retired Meter Reading Supervisor Carl Johnson. Rinard gave Berkeley support as the latter installed the Pit-Con filtration process at the Hayden Bridge Plant (previously described).

Ken Rinard employed a quiet, businesslike manner in approaching management problems. In his capacity as director of engineering and operations, he was by far the most knowledgeable employee in the technical aspects of the organization. He retired June 30, 1985.

1969; (left to right) Henry F. Beistel, director of administration; Charles H. Strong, chief engineer; Herbert H. Hunt, director of engineering and operations; and John E. Brown, treasurer and director of finance and accounting. (EWEB files)

Today, the water department continues under the capable superintendency of Kimber Johnson with Ronald L. Wilson assisting. Upon Rinard's retirement, Jim Brown was promoted to the position of director of engineering and operations.

(Left to right) Tom Davis, substation maintenance and construction; Lynn H. Baxter, electric department; Clark Lynch, steam plant; Leslie A. Pearson, safety. (EWEB files)

PERFORMANCE CRITERIA DEVELOPED

Harold H. King joined the organization in August, 1926. He left during the period from 1943 to 1946 to practice public accounting. Upon returning, he worked until his retirement in 1970. King developed criteria by which organization performance could be measured. The information he generated proved to be valuable to managers as well as the Board. As controller, although not in direct line of supervisory authority, King provided important assistance and consultation to Ralph Patterson, deputy director—finance and accounting.

Ralph B. Patterson was with the organization from March, 1933 to May, 1965. A meticulous fiscal manager, he guided the accounting department from its days of primitive accounting methods to the era of computers. When the IBM 1401 computer was being placed in service in the mid-1960s, as might be expected, there was recalcitrance among some longtime employees. Patterson worked to overcome this resistance and, along with Harold King, gave valuable assistance to Curtis

1962; (left to right) Supervisor Ray Nevers, Lawrence Fox, Jan Hoogstad, meter reading. (EWEB files)

Sprecher who assumed much of the responsibility for the computer installation.

Patterson practiced a management style which paid dividends to his department and certainly to himself. Each morning, almost without fail, he would visit his employees at their desks, briefly talking to them. If he was aware of family problems or illness, he would make appropriate inquiry.

For many years, Patterson supervised the meter reading, warehouse and credit activities. He was sensitive to the problems in these areas and some of the basic credit philosophies developed by Patterson and Credit Manager Earl Ritter are being carried forward today by Steve Hill, who supervises the credit activity. Fred Stuhr continued as warehouse supervisor for many years following Patterson's retirement.

A. Ray Barlow joined the operation in July, 1947 and for many years, served as cash accounting supervisor. Barlow was always a true gentleman. He never lost his patience nor genuine concern when dealing with employees who thought they might have been wronged on a paycheck. Barlow retired in 1973.

Henry F. Beistel joined EWEB in August, 1955. If Board member Egbert S. Wengert represented the study and theory of public service, Beistel represented the practice. He fell heir to many human-related activities including personnel and public relations, legislative affairs and relations with various units of local government.

1969; (left to right) electrical engineers Francis Whitney and Jim Howell; civil engineer Dean Hodges; Chief Draftsman Dean Wentworth. (EWEB files)

1968; (left to right) Public Information Manager Jack Smutz; Assistant Treasurer William F. Rau; Personnel Manager Norm Stone. (EWEB files, H. Gross photo)

CONCEPTS CARRIED FORTH

Beistel authored the first comprehensive employee-training program, elements of which remain in place today. Under his supervision, a pioneering concept in group-medical insurance was adopted December 1, 1956. He, in large part, was responsible for bringing about a modern personnel-relations system. Many of the concepts he recommended were carried forth by Personnel Services Manager Wes Goodrich and present-day incumbent, Dick Anderson. Similarly, in purchasing, Beistel worked closely with Purchasing Agent Frank Asay and procedures they developed have been continued and improved upon by Charles P. "Pat" Mulloy in recent times. Jack Smutz carried on in media relations and Tom Santee handles this responsibility today.

Much of Beistel's legacy remains in evidence. The training function, first under Judy Pierce and currently under Edmund Sheridan, includes concepts that Beistel set down nearly 30 years ago. The safety activity has gone forth. Donald M. Millar was the first safety person, followed by Leslie A. Pearson, John Brassard and Dale "Andy" Gump with today's safety activity under the direction of Elliott Lewis. Pearson worked in a broad administrative area as well as in safety. Administration of group insurance benefits has been ably carried forward by Nicole Gauron, Kathy Hudson, Kay Geiss, Ethelyn Brown and Ruth Dellinger.

Mel Damewood, electrical distribution engineer for many years, appointed assistant to General Manager Jean Reeder August 1, 1985. (EWEB files)

Beistel would have approved and welcomed the establishment of the regular positions of legislative representative and governmental affairs coordinator, occupied respectively today by Libby Henry and Debborah Brewer. He also would have quickly grasped the changing milieu of the utility industry and would have approved of the work being done by Energy Education Consultant Al Hughes and special consultant and energy-conservation columnist Val Thoenig. Some of Hughes' activities are an extension of concepts developed by Bill Eaton who was in Beistel's department for several years.

Beistel retired in June, 1971.

Curtis Sprecher joined EWEB in September, 1950 and retired in June, 1984. After working in various capacities in the accounting and finance department, he assumed a major responsibility for the installation of the first computer equipment in the mid-1960s. In this capacity he put in extremely long working hours addressing the many problems which were confronted.

Sprecher later became systems manager in which he worked as a consultant prioritizing projects scheduled for computerization. After the time of the original computer installation, many advances were made. Sprecher was an important factor in this process.

Without formal college education, Sprecher utilized correspondence training during his early EWEB years. In addition to this preparation, he was tireless in addressing his tasks.

Electrical engineers John Scofield (left); Dean Axtell (right). (EWEB files)

Carley J. Lewis joined the organization in October, 1951. Her career, still in progress, has been in the tradition of Grace Shaffner. She has capably performed many functions in the accounting department and in recent years has devoted much of her time to the payroll activity.

Lewis' father, Vern Gillespie, was employed for many years as an equipment operator on the upriver crew. Two of her uncles, George M. Easton (previously discussed) and Edward Howells, were associated with hydroelectric activity at Walterville and Leaburg.

(Left) Kenneth W. Rinard; (right) Ben A. Barnts.(EWEB files)

(Left) Richard D. Gall; (right) Carley J. Lewis. (EWEB files)

(Left) Grace Shaffner; (right) George B. Partridge. (EWEB files)

(Left) George Easton; (right) Lucille Norton (EWEB files)

GIVEN A JOB, IT WAS DONE AND DONE WELL

George B. Partridge's career at EWEB was in the tradition of his father, Burt Partridge as well as Dick Hart and Gus Parker. He joined the organization in November, 1948 and following a brief absence in 1956, returned to work until his retirement in January, 1982.

Partridge, after working on construction crews as an equipment operator for many years, eventually advanced to service operations superintendent—a position involving supervision of the shop and general construction activity. It was in this role that he proved his mettle and worth to the organization. When Partridge was given a job to do, it was done and done well. Notable among those assisting Partridge were Gene Brown, Wallace Kraegenbrink, Lee Eaton, Ray Davis, Wayne Franklin, Ralph Collins, Ray Wood, Ed Jordan, Del Smith, George Golden, Hans Skurdol, Jimmy Harrison, Gary Winters, LeRoy Lueck and Emmett "Fuzz" Lovell.

Various employees represented the organization well in capacities away from EWEB. Governor Bob Straub appointed Treasurer and Director—Finance and Accounting John E. Brown to membership on the Oregon Debt Advisory Council. Brown, who served four years on this body, was its chairman for 3½ years. He was one of the first appointees to the council which was formed during Straub's administration. The council was instrumental in securing an increased bond interest rate ceiling—an important factor in Oregon municipal-bond financing.

(Left) Glenn Haney; (right) Kimber Johnson. (EWEB files)

(Left) Barbara Lamb; (right) John Fuller, both valuable employees in power resources. (EWEB files)

Francis Whitney served and Owen Brown serves today as a member of the State Board of Engineering Examiners. In the area of industry education Lloyd Lindley, for many years, taught basic electricity and other subjects to electric-line apprentices at Lane Community College and still serves as a consultant to that institution. Albert Kronberger taught electric metering at the University of Washington.

The employees mentioned in this chapter are, in the writer's view, a representation of the EWEB work force. Hundreds of valuable employees have not been specifically named in these pages. This does not diminish their importance.

This chapter is intended as a statement of the character and dedication of a community of employees with an essentially common purpose.

CHAPTER ELEVEN
Other Perspectives

"PLEASE GIVE MY warmest regards to everyone there at EWEB; tell them 'Merry Christmas and Happy New Year'; I have very fond memories of Eugene and EWEB."

These were the words of Cecil Andrus on December 22, 1976. At the time Andrus was governor of Idaho and President-elect Jimmy Carter's choice for secretary of interior.

Andrus worked for the Eugene Water Board during the summer of 1948 after graduating from Eugene High School. He also worked the following summer and after spending some time at Oregon State, became a full-time employee for a year and a half during 1950-51.

In a telephone conversation with EWEB Intercom (in-house bulletin) Editor Glen Stadler and Electric Distribution Engineer Mel Damewood, Andrus said, "Before you ask me any questions, don't believe a thing those guys at EWEB tell you—like Glenn Haney, Vern Poindexter, Don Woodward, Willie Dotson, Clark Lynch, Doug Frye, Pete Humphrey, Loyd Coons, Darrell DeLong, Bud Frye, Lou Polley, George Partridge, Jess Trunnell, Ike Lusk and all my other friends there."

Stadler said, "'There's the story about your last day on the job." This alluded to the report that Andrus, when he had completed his final job for Glenn Haney, punted his lunch bucket over the fence and into the Willamette River.

". . . when I kicked my lunch bucket over the fence and into the river?" the governor interrupted. "That's true, all right. But, being a frugal guy, I took my thermos out first." With the statement, "I'll be darned if I'm gonna dig ditches all my life," Andrus left EWEB for the Navy and the Korean conflict.

A bow-hunter, Andrus spent considerable time with Damewood stalking deer in the woods. He also was on the Water Board softball team, playing pitcher and second base.

"A REAL 'SELF-STARTER'"

Glenn Haney said on the day of the phone conversation, "Cece was a real 'self-starter.' He'd come to work and right away grab a shovel or some tools and start right in, while some of the other fellows would wait until I assigned them a job. Cece was smart that way. He knew what had to be done, and did it. Of course, because he was already doing something, more or less of his own choice, he didn't get some of the assignments he might otherwise have gotten."

Following service in the Navy, Andrus and his bride, Carol May of Eugene, moved to Idaho. He served in the Idaho state senate, was defeated for the governorship in 1966, but won in 1970, the same year Jimmy Carter won on his second try for governor of Georgia.

"MEET NORM STONE FROM BRAND X"

During the early 1970s, at a meeting in Seattle dealing with President Nixon's economic recovery program, the writer sat at a table with several other utility industrial relations officials. Among them was Wallace E. Gordon, then vice president of industrial relations at Portland General Electric Company.

Gordon, good naturedly, introduced the writer as being from "Brand X in Eugene."

Why such an introduction? The immediate answer had to do with EWEB's nonunion status and long involvement in the Northwest utilities wage and hour committee. EWEB was and is the only member of this organization without a union contract. This nonunion condition engendered the envy of other utility representatives who were battle-worn veterans of perennial contract negotiations. Gordon had more than a cursory knowledge of the Eugene utility operation. He had provided valuable assistance and advice when EWEB formalized its personnel program in 1956.

But there was more to the "Brand X" designation than nonunion status. In September, 1984 Robert H. Short, chairman and chief executive officer of Portland General Electric Company, provided an excellent written commentary about his employment as editorial assistant at EWEB from February, 1953 to April, 1955.

"THE TRANSITION YEARS"

Referring to the 1953-55 period as "the transition years," Short wrote, "The year 1953 marked the beginning of what was to become a significant change in the attitude and policies of the Eugene Water & Electric

Board—as an electric utility, as a public agency actively participating in community and legislative matters and as a governing board.

"Historically, EWEB was conservative in virtually all of its practices. It did not take part in the public power efforts to acquire the properties of Oregon's investor-owned electric utilities which took place in the late 1930s and early 1940s. In fact, although it was neutral in its public posture, privately it supported the position of the investor-owned companies. The Board itself was strongly business-oriented and considered itself more philosophically aligned with the investor-owned utilities than with the advocates of public ownership."

Short is correct in his assessment of the Eugene utility operation's political stance during the height of the public power movement in the 1930s and 40s. However, it should be remembered that in early times, the Eugene Water Board was very aggressive in its takeover of the old Oregon Power Company.

Short continues, "It was not until 1953 that EWEB had any consistent representation at the state legislative session held biennially in Salem. Neither did it take part in city government even though it was a creature of the Eugene City Council. Prior to 1953, EWEB had no public

(Left) Cecil Andrus, former governor of Idaho and secretary of interior in President Jimmy Carter's administration; (right) Robert H. Short, chairman and chief executive officer, Portland General Electric Company. (Yuen Lui photo)

relations or public information department. It did take part in the meetings of the Northwest Public Power Association but only in the technical sessions, staying aloof from the politics of public power.

"In short, EWEB considered itself an electric utility, conservatively managed, with little interest in external affairs, with limited ambitions to expand its service area beyond the Eugene city limits, and with virtually no interest in political activity."

Short says in his commentary that it was through the efforts of himself, Board member Stan Darling and Byron Price that the traditional and conservative attitudes began to change. Price joined the organization as assistant to General Superintendent-Secretary Ray Boals in July, 1953.

PROFESSIONAL LOBBYIST EMPLOYED

The first indication that change was taking place occurred in January, 1953 when the Board decided to employ a professional lobbyist to represent its interests at the state Legislature. The individual was Raymond M. Kell, a Portland attorney active in the politics of public power. Short comments, "Not only was the employment of a lobbyist an indication of changing attitudes, but the employment of a public-power activist was unheard of."

Hiring a lobbyist was a major step, but the most significant indication of change were events in early 1953 which ultimately led to the acquisition of properties owned by a private utility. Short writes in his commentary, "Largely in response to high electric bills rendered by the Mountain States Power Company, an investor-owned electric utility which served much of the area adjacent to the city of Eugene, a number of residents in a suburban area north of the city petitioned EWEB for electric service. Normally, requests such as this would have been handled with a polite 'no,' but in this instance Board member Darling insisted that the petitions be taken seriously and that the Board employ Ray Kell to advise the Board as to 'what procedure is to be followed by the Board in dealing with the Mountain States Power Company for acquisition of (its) properties now in the areas in which petitions have been received by the Board requesting our service.'"

Following guidelines established by Kell, acquisition of the property went forward. In June, 1953 notification of EWEB's intent to serve was sent to Mountain States Power Company and to petitioners in the designated area. Short writes, "In July, 1953 EWEB had built into the Mountain States Power Company territory and begun connecting Mountain States Power Company customers. EWEB crews would contact a (customer), obtain written approval to serve, remove the Mountain States meter, install an EWEB meter and connect the customer. At the end of the first day's action, EWEB staff called Mountain States staff in the district office and advised that, 'we have 17 of your meters in our storeroom. Will you please come and pick them up.'

"COMPLETE SURPRISE"

"This uncharacteristic and aggressive competitive activity came as a complete surprise to Mountain States Power Company according to Ray Kell who, at a special Board meeting on July 27, 1953, reported on his meeting with Mountain States officials earlier that month. The Mountain States Power Company indicated no willingness to discuss the matter, so the Board, now with the competitive bit in its mouth, authorized staff to build a duplicate distribution system in the area and connect anyone who wanted service, petitioner or not, with the goal being 75-100 customers per month."

Short comments further, "It should be noted here that EWEB's activity was not universally supported within the organization. The superintendent and most of the senior staff were reluctant participants. The strategists were Darling, Kell, Short and beginning in July, Byron Price. It also should be noted that Mountain States Power Company's 'complete surprise' was totally justified. That conservative EWEB would aggressively pirate customers of an investor-owned utility was incredible.

"At its August, 1953 meeting, the Board dropped the other shoe; it authorized appraisal of all Mountain States Power Company property in areas 'being considered for inclusion in the EWEB system.' Up to this point EWEB had simply been taking customers; now it was considering taking all of Mountain States Power Company's property.

"Talks with Mountain States produced nothing, so on December 14, 1953, the Board approved a resolution to acquire by eminent domain (condemnation) the property in question.

"During the ensuing months EWEB continued to connect Mountain States customers. Mountain States did nothing, principally because it was in the process of being acquired by Pacific Power & Light Company, a multi state Pacific Northwest investor-owned utility.

"On July 7, 1954, the Board approved arrangements for a meeting that night with PP&L to discuss purchase of the former Mountain States property. The meeting produced no tangible results but later that month, after several sessions between Darling, Kell, and PP&L counsel, Kell reported that PP&L would like to sell, but for tax purposes it could not be a willing seller; the properties must be condemned.

"Approval to acquire the properties by formal condemnation was granted by the Board on August 27, 1954. The price of $1.22 million was set on August 31, 1954. (Fittingly enough, both actions were taken on the motion of Stanley Darling). The amount was derived by using a multiplier of annual revenue from the area to be acquired. In this case it was

5.6 times, unusually high, but the Board had its mind made up and was not about to quibble over numbers.

"The condemnation order was issued by the Lane County Circuit Court on August 30, 1954 and the last PP&L customer was cut over to the EWEB system on September 27. An eventful chapter in EWEB history had come to an end."

Short also became involved in a legislative effort to impose ad valorum taxes on EWEB property outside the Eugene city limits. During the third week of March, 1953, a bill was introduced whose sole beneficiary was Lane County School District 19 (Springfield) "in which were located all of EWEB's hydroelectric plants and a significant part of its transmission system."

Short and Ray Kell were assigned the task of lobbying against the bill. Introduced by a Republican in a Legislature heavily dominated by that party, it appeared likely that the bill would pass.

Kell, a Democrat, elected to limit himself to strategy planning with Short taking the role as visible EWEB representative. In the ensuing process Short contacted as many legislators as he could. His simple, straightforward message was, "that EWEB agreed that public-power agencies should help support public services in areas where there were facilities owned by public-power agencies, but the bill being considered was flawed. EWEB would undertake a major study of the entire question of taxation of public power and submit recommendations to the 1955 Legislature. Further, the current proposal addressed only the EWEB-School District 19 situation. Clearly, the issue was far broader than that and should be resolved on a statewide basis."

The strategy worked. The legislation was turned aside and the Board, keeping its word, adopted a resolution authorizing a comprehensive study of the entire question of public-power agency taxation.

The 1957 session passed legislation permitting, not mandating, in lieu of tax payments to school districts. This entire process culminated in 1959 when Byron Price, early in his administration, made a comprehensive recommendation to the Board which included substantial payments to school districts in which EWEB facilities were located. EWEB was well on its way to becoming a major "taxpayer" in Lane County.

BOARD ENTERS NATIONAL POLITICS

Short describes EWEB's first overture into the arena of national politics. "Early in the first Eisenhower administration, the U.S. Interior Department began formulating a policy which would reduce the federal government's financial involvement in water resources development. The concept was that in the case of (the) construction of multiple-purpose water resource projects, the United States would finance the portions allocated to public purposes, i.e., flood control, navigation and irrigation, and nonfederal agencies would be permitted to finance and receive the output of the power-generating facilities.

"Long before July, 1953, the U.S. Corps of Army Engineers had developed plans for construction of a flood control and power-generating project on the south fork of the McKenzie River some 40 miles east of Eugene, designated as the Cougar Project.

"The Board had also been considering the development of hydroelectric power on the Upper-McKenzie River. At its July 13, 1953 meeting, EWEB staff suggested that (the Board) examine whether the Cougar project could be economically tied in with EWEB's plans for power development on the McKenzie. The Board concurred and authorized staff to retain the Corvallis engineering firm of Cornell, Howland, Hayes and Merryfield to examine the concept.

"From July to December, 1953, the McKenzie studies continued. On December 28, 1953, the Board authorized staff to prepare the material necessary to present the Cougar 'partnership' proposal to Congress. The

Trojan Nuclear Project. (Portland General Electric files)

plan was for EWEB to finance—but not own—the power-generating facilities (50,000 KW) at the project and receive the output for the life of the project. Since Cougar was an authorized federal facility, legislation would be required to permit nonfederal participation.

"At the same meeting, the Board authorized staff to file for a Federal Power Commission license to construct the Beaver Marsh project on the upper-McKenzie River."

It was Board member Darling, Byron Price and Short who conceived the Cougar proposal which was introduced in the Congress by Representative Harris Ellsworth, a Roseburg Republican, in February, 1954. The task of lobbying the bill, HR 7851, was given to Price and Short, with assistance from Ray Kell and Thomas B. "Burke" Hayes of the Corvallis engineering firm.

At a hearing of the House Public Works Committee on April 25, 1954 Price and Short had arranged for the appearance of an impressive list of supporting witnesses. Present were Oregon Governor Paul Patterson, Oregon's senior senator, Guy Cordon, Congressman Ellsworth, General Emerson C. Itschner, chief of the U.S. Army Corps of Engineers (later a vice president at Portland General Electric Company) and Thomas W. Delzell, then chairman of the board at PGE.

Short continues, "The bill was reported out of committee with a favorable recommendation and came before the full House early in May, 1954. In almost a perfunctory proceeding, the bill was approved by voice vote.

OPPOSITION ENCOUNTERED

"The Senate was an entirely different matter. By the time the bill reached that body, it had attracted the attention of militant and traditional public-power supporters. Unlike EWEB and a few other municipal utilities, most public-power agencies depended on the largesse of the federal government for survival. Power generated at federal multiple-purpose water resource projects was sold to them on a preferential basis. They feared that the 'partnership' concept would, in the future, deny them access to low-priced federally generated electric power, without which they would lose any competitive advantage over investor-owned utilities.

"Leading the opposition against HR 7815 was Oregon's junior senator, Wayne Morse from Eugene, who earlier had changed his party affiliation from Republican to Independent. Allied with Morse were Washington's two Democratic senators, Warren Magnuson and Henry M. "Scoop" Jackson.

Early in June, 1954 Short was dispatched to Washington to oversee the Cougar bill through the Senate. With the aid of Senator Cordon, a Senate committee meeting was scheduled and held. This resulted in a favorable committee report and the bill was scheduled for action by the full Senate late in July.

Continuing the account, Short says, "On the day prior to floor consideration, Senator John Pastore, Democrat from Rhode Island, commenced a filibuster against a provision of the Atomic Energy Act which he considered unfair to public-power agencies. The filibuster lasted almost two weeks. When it ended, Senator Cordon sought to bring the Cougar bill to the floor for Senate action. Senator Morse announced that he would filibuster against HR 7815 'for as long as it is necessary.'"

The situation in August, 1954 was bleak for Short and the Cougar legislation. Washington, D.C. in 1954 was not air-conditioned and in August, the city was intolerably hot and humid; everyone wanted to adjourn and go home. Short observes, "It was an election year and those senators up for reelection (including Senator Cordon) were anxious to return to the hustings. The prospects of spending the remaining days of August suffering through a Wayne Morse filibuster were unattractive indeed.

FIRST VENTURE UNSUCCESSFUL

"Senator Cordon was persuaded by his colleagues to remove the Cougar bill from the Senate calendar. EWEB's first venture into the arena of national affairs did not succeed."

Governor Julius L. Meier, ran successfully on a public-power platform in 1930.

Short's concluding comments relate to the "Brand X" designation alluded to in earlier pages. He says, "During the Cougar debate, EWEB found itself in a curious position. At home it was taking over part of an investor-owned utility's area, an action heartily supported by public-power groups. In Washington, D.C. it was in a leadership position in a legislative effort bitterly opposed by those same groups.

"Although this might have seemed contradictory, in actuality it was a demonstration of the Board's independence. EWEB was in the process of acquiring Mountain States Power Company's territory because it was a good revenue producer; public-power advocacy had nothing to do with it. The Board embarked upon the Cougar effort because it would have provided a source of low-cost hydroelectric power; politics were never considered.

"One unintended result of the Cougar effort was the development of a long-standing relationship between EWEB and Portland General Electric Company. PGE supported EWEB in the Cougar proposal and EWEB agreed in mid-1954 to take part in a similar and equally unsuccessful effort with PGE on the John Day project."

Short speaks of the personal relationship between himself and Byron Price that "would last long beyond our days at EWEB. This personal relationship was probably the most significant factor in EWEB's decision to take an ownership position in PGE's Trojan Nuclear Project without which the plant would likely not have been built. It also provided the foundation for a mutually respectful and supportive relationship between the staffs of the two organizations which exists today."

During the early 1960s, the writer had the opportunity to spend an hour with Bob Short when the latter was a PGE vice president. The purpose of the visit was to obtain views on EWEB's organizational structure that Short might have developed during his EWEB employment. Those views were transmitted to Byron Price and it is the writer's belief that they proved helpful as Price pursued the management of EWEB.

Is there irony in the fact that Bob Short, who was a major factor in a public-power utility acquisition of investor-owned utility territory, would later become chairman and chief executive officer of another investor-owned utility? Not if it is kept in mind that Short is an achiever —one who excels in whatever he pursues.

OREGON OPPOSITE OF WASHINGTON

When Governor Bob Straub's Domestic and Rural Power Authority (DRPA) bill was being considered, it was observed that while some 80 percent of residential customers in the state of Washington were served by public power, only 20 percent of such customers in Oregon were served by public agencies. Related to this was the fact that after both states enacted PUD legislation in 1931, 25 utility districts were created in Washington and only five in Oregon. Thus a far greater proportion of Washington's domestic customers were beneficiaries of preference power—Bonneville power on which public agencies had first call. (The disparity between the two states has been largely corrected by passage of the Regional Energy Act which allows investor-owned utilities to acquire Bonneville power at the preference rate for their rural and residential customers.)

Why the difference? There certainly was no lack of public-power fervor in Oregon in the early 1930s. Julius L. Meier, during his successful campaign for governor in 1930, said, "The paramount issue in this campaign is public power development versus private development. This issue is the very heart and soul of the battle for progressive principles. Which shall it be—private development for the further enrichment of the coffers of the power trust, or public development for the benefit of the state of Oregon and its people?"

Ken Billington, for many years executive director of the Washington Public Utility Districts Association, offers an answer as to the difference in the two states. Quoted in Stephen A. Stone's *Oregon Memorabilia* in 1967, Billington says, "The answer is very simple. The Oregon PUD law has a restrictive provision on self-financing placed in it by the private-power companies. In the state of Washington revenue bonds used to finance PUDs, and are not mortgages or liens against the utility properties, can be issued under the same kind of business decisions and judgement as those exercised by a board of directors of a private corporation. Conversely, in Oregon, issuance of revenue bonds is subject to political decision. It is masqueraded under the high-sounding principle of 'letting the people vote on bond issuance.'

"Here you have substituted political decision for business decision. The private power companies, with an unlimited and uncontrolled budget for political purposes, can easily stir up enough confusion or controversy over the issuance of revenue bonds so that it is impossible to purchase or take over their properties. People decide by emotion rather than judgement."

Others say that historically, PUDs were not needed in Oregon because the investor-owned utilities cooperated by building rural line extensions and keeping rates low.

Billington and Gus Norwood, executive director of the Northwest Public Power Association for many years, were astute and effective public-power advocates. At public-power meetings, they sometimes gave a presentation billed as the "Billington-Norwood Show," a public-power industry version of the "Ev and Charlie Show," the frequently held press conferences during the 1960s featuring Illinois Senator Everett Dirkson and Indiana Congressman Charles Halleck.

It has been the writer's observation over many years that electric utility customers have been the true beneficiaries of the public power-private power struggle in a region where battle lines are more sharply drawn than in most other areas. It might be because of the "check and balance" condition which has existed. The major support for this observation lies in the incredible disparity in electric rates between the Pacific Northwest and other areas of the country. The region, to be sure, has enjoyed the benfits of a hydroelectric generating base. But low-cost hydroelectric power notwithstanding, electric rates in other areas appear to be dramatically and unjustifiably high.

CHAPTER TWELVE
Looking Ahead

IN HIS "REMINISCENCES" J. W. McArthur attributes the following to Mayor Joseph D. Matlock. "In my mind, water is associated with health, light with learning, and it is our aim that each child, however poor or obscure, shall have at the minimum cost these two fundamentals of civilization."

Matlock, W. W. Calkins and Martin Svarverud were men of vision. In their time and circumstance they acted to establish a community service that would endure well beyond the world they perceived.

In every community, including Eugene, there are those who work, not for acclaim, but for the betterment of their world and the world which belongs to posterity. Monsignor E. J. Murnane, Les Anderson, Maurie Jacobs, Ted Baker, Ruth Bascom, "Windy" Atkinson, Betty Niven, Ed Cone, John Alvord and Dick McDuffie, longtime Weyerhaeuser official, are but a few of those who have worked, often quietly, for a better community and whose contribution will endure long past their perceived world.

So what of the future of a particular aspect of this community—its municipal utility operation? With a fine Board of Commissioners and a general manager, Jean Reeder, who has the capability of performing in an excellent manner, the operation should possess the confidence necessary to continue its record of outstanding performance and to move ahead with new and innovative ideas.

Everett Jordan, EWEB generation manager, served aboard a nuclear-powered submarine during his years in the Navy. Speaking of cruising submerged for 90 days at a time, Jordan said recently that upon resurfacing after a lengthy stay under water, one perceives that the world has changed—that it is somehow different.

Jordan's observation is a practical example of some of the ideas of Georg Wilhelm Friedrich Hegel. Hegel, an early-19th century German philosopher, said that change is the fundamental condition of human activity. He tended to politicize his own philosophy and was criticized for this by some of his peers. Nevertheless, one of Hegel's ideas merits mention.

Hegel said that in the process of any human activity, there is always a thesis. By its very nature, the thesis creates an opposite or antithesis. The thesis and antithesis interact to form a synthesis and the synthesis then becomes a new thesis and the process continues.

Whether looking to Jordan's astute observation or studying the philosophy of Hegel, the important idea is that at any point in time, things will never be what they were before.

Over many years EWEB has tended to function in a light-cavalry mode. It has avoided the pitfall of becoming a sluggish bureaucracy. It has moved quickly and decisively when required to do so in the interests of its customers. And it seems to have understood the dynamics of change. If there is admonishment at all in these pages, it is that the tradition of decisiveness, understanding of change and rapid response to problems continues.

Not as admonishment, but as suggestion, EWEB commissioners and management might consider studying additional opportunities for interutility and intergovernmental cooperation. There is a long-history of such cooperation including sales to water districts, power-purchase and exchange agreements, water storage arrangements, loaning crews in time of wide-spread outage, providing billing service to other entities and for a period, sharing safety services with other utilities. Many years ago the Utility Coordination Council was established. EWEB's Mel Damewood was instrumental in this formation and the council has since become a state-wide activity which has yielded dividends in terms of savings to ratepayers, safety and prevention of dig-ins. Currently, EWEB is involved in a joint-agency common-mapping project with Lane County, City of Eugene and Lane Council of Governments. Chief Draftsman Ron Davis is taking a leadership role in this process.

In years past EWEB staff discussed the seemingly redundant condition existing locally in which several utilities maintained a concomitant number of equipment pools. When EWEB Commissioners Calvin Schmidt and Taylor Ramsey served as a special committee during the 1970s to study EWEB construction practices, an important issue on their minds was the use-factor of various kinds of equipment. As entrepreneurs in the construction field, they were keenly aware, at least from their point of view, of the need to *use* equipment in order to realize a satisfactory rate of return on the investment required to purchase that equipment.

Kenneth C. Tollenaar, Director of the Bureau of Governmental Research at the University of Oregon, has observed and studied many examples of intergovernmental cooperation. His words that follow, although specifically addressing local government and intergovernmental relations, are germane to utilities and interutility cooperation. Even investor-owned utilities are quasi-governmental entities inasmuch as they are regulated and legally committed to meet a public need. In a paper prepared in 1983 Tollenaar summarizes, "Clearly, intergovernmental cooperation arrangements should be utilized if analysis

Jean Reeder, named general manager-secretary July 1, 1985. (EWEB files)

suggests that economies might be obtained for the local governments involved. It appears that many of the problems experienced under these arrangements might be avoided through careful planning and drafting of intergovernmental agreements. Various actions might be taken to improve utilization of intergovernmental arrangements, including development and dissemination of information on types of services and facilities available from various service providers; initiation of specific proposals by units willing to provide services intergovernmentally or by councils of government; and technical assistance to help local governments negotiate and formalize effective and workable intergovernmental agreements."

Neither Tollenaar nor Freeman Holmer, Eugene city councilor and longtime student and practitioner of government, view intergovernmental cooperation as an instant remedy that will bring about significant economies. Holmer said recently that a program of cooperation which may be superficially desirable might present many problems which must be resolved before satisfactory implementation.

The key, as Tollenaar points out, is a process of careful analysis and planning. He states in his written remarks that while research indicates that economies of scale are available in all governmental functions at some point, savings do not automatically derive from joint and cooperative efforts.

If EWEB were to consider expanding the concept of inter-agency cooperation, perhaps a useful model would be the intergovernmental services board which has been established by the Oregon cities of Monmouth, Dallas and Independence, and Polk County. This board has the responsibility of overseeing several existing joint and cooperative arrangements and exploring additional ways to cooperate.

As EWEB considers its master plan which ultimately will lead to new quarters, it should include provision for a utility museum-history activity. As the organization nears its 100 year anniversary, its history will become increasingly important. Due to the work of retired EWEBer Bill Eaton, the organization possesses a rich collection of artifacts associated with the utility industry and EWEB.

Beyond adequate funding, there are three requirements for this kind of program. Firstly, it must have the unqualified support of the Board and management. Secondly, adequate and appropriate space must be provided. Thirdly, appropriate personnel must be identified who will assume responsiblity for the program. Such personnel, in the tradition of Earle Clark and Bill Eaton, must have a sense of organization history and the importance of preserving it.

The Eugene Water & Electric Board, its management and its work force possess the skill and dedication to move the utility operation into the third millenium with the same, if not higher, level of success enjoyed by the organization during its first three-quarters of a century.

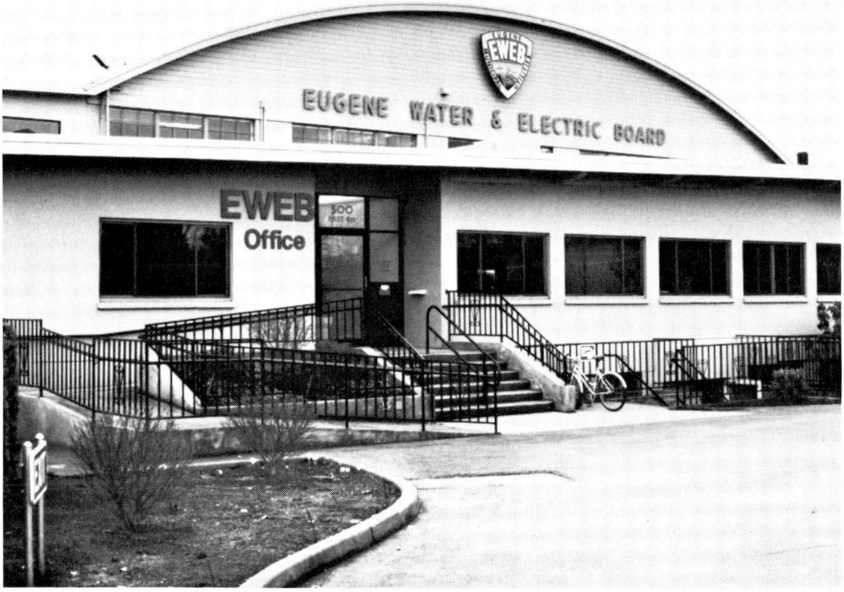

1986, EWEB Headquarters (Northwestern Photographics).

INDEX